Aldair dos Santos Gomes
Dihego Souza Pessoa
Viviane F. Silva

Water scarcity and agricultural production in semi-arid Paraiba

Aldair dos Santos Gomes
Dihego Souza Pessoa
Viviane F. Silva

Water scarcity and agricultural production in semi-arid Paraiba

Estimating agricultural production and analysing rainfall for water management

Imprint
Any brand names and product names mentioned in this book are subject to trademark, brand or patent protection and are trademarks or registered trademarks of their respective holders. The use of brand names, product names, common names, trade names, product descriptions etc. even without a particular marking in this work is in no way to be construed to mean that such names may be regarded as unrestricted in respect of trademark and brand protection legislation and could thus be used by anyone.

Cover image: www.ingimage.com

This book is a translation from the original published under ISBN 978-613-9-70338-8.

Publisher:
Sciencia Scripts
is a trademark of
Dodo Books Indian Ocean Ltd. and OmniScriptum S.R.L publishing group

120 High Road, East Finchley, London, N2 9ED, United Kingdom
Str. Armeneasca 28/1, office 1, Chisinau MD-2012, Republic of Moldova, Europe
Printed at: see last page
ISBN: 978-620-8-18835-1

SUMMARY

PRESENTATION

This manuscript was produced with the aim of adding knowledge about agricultural production and rainfall analysis in the most recent water crisis that occurred in the Northeast, particularly in the state of Paraíba. The prolonged drought that occurs frequently in these Brazilian regions is one of the factors that hinders agriculture and economic growth, especially in rural areas, where the water deficit encourages rural exodus.

Chapter 1 addresses the issue of agricultural production from 2008 to 2017, with temporary and permanent crops such as herbaceous cotton, peanuts, rice, beans, maize, bananas, potatoes, sugar cane, cashew nuts, tobacco, oranges, manioc and tomatoes. These are crops that are part of the population's daily life and that drive trade. Using Gretl software, it is possible to analyse by ordinary least squares what happens over the years and how much drought directly affects the agricultural area, since water is necessary for plant development.

Livestock farming is another branch applied in rural areas, but without water and food it is difficult to maintain the herd of animals, often in times of severe drought, the lack of water is a restriction on the type of animals to be raised as well as the crops to be planted. Chapter 2 looks at the amount of water needed for livestock farming, based on rainfall and local livestock farming in a municipality in Paraíba. Large animals generally consume more water and food, and for semi-arid regions, smaller animals are interesting in trying to overcome drought in the best possible way. There is no way to combat something that is natural, drought is a natural phenomenon that happens in the Brazilian Northeast and the more information you get about this region, the more possibilities there are for living together and the better the quality of life for those who live in these places.

In chapters 3, 4 and 5, an analysis of the rainfall that occurs in a given location makes it possible to analyse the occurrence of rainy or average seasons, as well as lower than expected rainfall. Through this rainfall analysis, management of the type of crops and livestock to be used can be planned, reducing losses and adapting to current conditions. Water harvesting is one of the most widely used alternatives, including the use of slab cisterns, which are a federal government incentive to mitigate social impacts.

The Boqueirão reservoir is responsible for supplying water to several cities in Paraíba and assessing rainfall in the municipality where it is located is important, although it is necessary to check all the tributaries that contribute to the volume. But analysing the local rainfall gives an idea of what is happening in other locations.

CHAPTER 1

ESTIMATING PLANTING AREA AND PRODUCTION IN TIMES OF WATER SCARCITY

Carlos Vaillan Castro de Bezerra, Dihego de Souza Pessoa, Aldair dos Santos Gomes, Julia Soares Pereira, Viviane Farias Silva & Vera Lucia Antunes de Lima

SUMMARY

The use of water in agricultural activities is a subject of great interest to those involved in this activity due to its high consumption, especially in irrigated agriculture. This is exacerbated in regions where the magnitude of losses *is* greater than that of rainfall, leading to a water deficit, as in the semi-arid region of Brazil, and climate monitoring is important to help manage agriculture. The deterioration of water resources is caused by various sources, from the unsustainable use of nature to the loss of species of fauna and flora that are important to the ecosystem. This research was carried out in order to assess the area and agricultural production in times of drought in the semi-arid region of Paraíba, using the Municipal Agricultural Production (PAM) research database from 2008 to 2017. The areas planted and the quantity produced of 15 permanent and temporary crops in the region were evaluated. The data was processed with Gretl 2018a software, using ordinary least squares (OLS). The results showed that increasing the cultivated area led to an increase in production rates, but this was due to other factors that directly influence the crops, since they depend on water for irrigation;

Keywords: Irrigation, agriculture, climate monitoring, water deficit.

INTRODUCTION

Water is a renewable natural resource that is important for all living beings and for maintaining life on Earth. When considered as a renewable resource, it is implied that it is unlimited and that in its cycle it returns to the Earth's surface. What happens *is* that the liquid surface is evaporated by the action of solar heating, the greater the heating the greater the rate of evaporation, plants lose water to the atmosphere through transpiration, the particles of water that pass from the liquid to the gaseous state form clouds that are carried by the wind to another location and when they are charged, the precipitation of water droplets falls by the force of gravity and with the pollution of the air, soil and water reservoirs, the water that returns to the surface becomes contaminated. Thus, despite the renewable nature of environmental pollution, it de-characterises the purity of the water *that is* returned to the surface, limiting the quality of the water, even if it doesn't change the quantity. Olivo and Ishiki (2014), reporting on the scarcity of Brazilian water, point out that the lack of water is a climatic adversity and the relevance of water to socio-economic sectors, which need to be addressed by specific laws.

According to Venancio et al. (2015), the deterioration of water resources is caused by various sources, ranging from the unsustainable use of nature to the loss of species of fauna and flora that are important to the ecosystem, and the prospect of a society that seeks environmental sustainability integrated with an economy that is abundant but respectful of nature is fundamental.

In some regions, such as the south and north of Brazil, there are perennial rivers with an abundance of water, such as the Amazon River. In the Northeast, however, the rivers are intermittent, with scarce and uneven precipitation, as well as long periods of drought that contribute to rural exodus and socioeconomic and environmental decay. Silva and Alcantâra (2009), studying space-time variability, reported that when the El Nino occurs, the number of days with droughts is longer compared to La Nina, with approximately 90 days without precipitation, while in the La Nina period this period is up to 40 days, affecting crops. These authors state that Pernambuco and Paraíba were the states most affected, with the longest periods and the largest areas affected.

As agriculture needs water for irrigation, it is common knowledge that the sector that uses the most water is irrigated agriculture. If crops are subjected to a water deficit, this leads to a reduction in production and even the loss of the entire plant. In the interior of Paraíba, one of the states that forms part of the Brazilian semi-arid region, an area planted with coconut palms, with the severe drought that has occurred in recent years, has extinguished this crop, and this has not only happened to the coconut palm, there have been several plantations that have been unable to withstand the long period without water, either from irrigation or from rain. With below-average rainfall for a number of years, reservoirs dried up, farmers had no grain in storage or lost out trying to grow crops just when it looked like the water situation would be eased, but the rainfall came quickly, with lower amounts, and then the sun remained constant, losing the entire cultivated area. With the reservoirs restricted only to the consumption of the population and animals, irrigation could not be applied to agriculture, while many people in mainly rural areas were supplied by water tankers when they arrived.

Agriculture is directly affected by long periods of drought, which is why it is important to apply climate monitoring to help manage agriculture, such as the favourable time to harvest, what type of crop to use, if the rainfall is for short periods it is desirable that the cycle is short, reducing crop and economic losses, becoming an instrument so that in difficult times the impacts are mitigated and irrigation is carried out correctly in the quantity and time that the crop needs are met (Silva and Alcântara, 2009). The fight against water waste in agriculture is verified in scientific research that seeks efficiency linked to a high production rate, since high amounts of water in irrigation cause soil and nutrient losses and crop losses. The amount of water needed for plants to develop as proposed must be achieved at different phenological times.

Water restriction is one of the main factors for agricultural production, but analysing the agricultural production of the semi-arid state based on the rainfall that occurred throughout the region is not coherent due to the uneven distribution, which is a predominant aspect of the semi-arid region. The area destined for agricultural production in times of drought is reduced, as most crops need water for their development. As water scarcity has become more significant over the years, analysing production over time is more comprehensive and less likely to result in errors when carrying out statistical analyses.

This research was carried out to assess the area and agricultural production during the dry season in the semi-arid region of Paraiba.

MATERIAL AND METHODS

The research was carried out in the state of Paraíba (Figure 1) using the Municipal Agricultural Production (PAM) research database for the years 2008 to 2017, available on the IBGE's automatic retrieval system - SIDRA (IBGE, 2018).

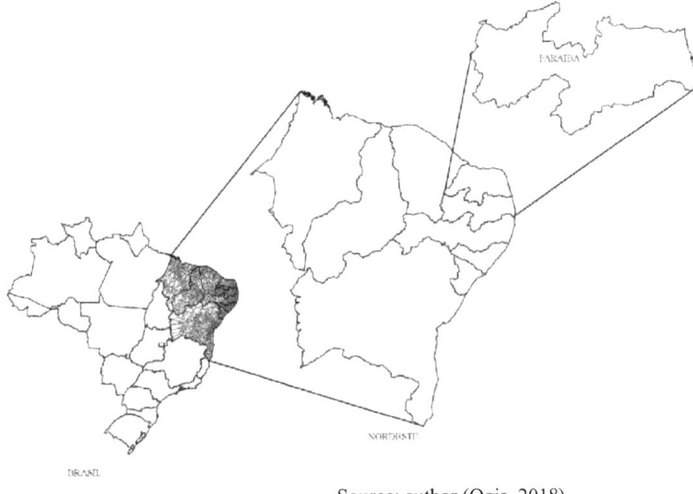

Source: author (Qgis, 2018)

Figure 1 - Geographical location of the Northeast region and Paraíba, Brazil.

The area planted (hectares) and the quantity produced (tonnes) of the following permanent and temporary crops were assessed: herbaceous cotton, peanuts, rice, beans, maize, bananas and potatoes,
sugar cane, cashew nuts, tobacco, oranges, cassava and tomatoes. The data was processed with Gretl 2018 software, using ordinary least squares (OLS).

RESULTS AND DISCUSSION

- Herbaceous cotton

By estimating the planting area in relation to the production of herbaceous

7

cotton, Table 1 shows that the dependent and explanatory variable justify around 81%, according to the r-squared. Thus, approximately 19% of the oscillations that occur may be related to other variables that are not included in the evaluation, such as rainfall. Production is statistically significant at the 1% level.

Table 1. Ordinary Least Squares, with the dependent variable herbaceous cotton production in the state of Paraíba from January 2014 to December 2017.

	Coefficient	p-value
Constant	-1,31685	0,8575
Area (A)	1,14225	7,43e-018***
R-squared	0,810334	

**Significant at 5% ***Significant at 1%

Note: data subjected to first difference to become stationary.

According to Table 1, increasing the area under cultivation by 1 hectare will add 1.14 tonnes of cotton to production. It can be seen that the amount of herbaceous cotton produced is related to the area under cultivation, so reducing the area under cultivation will cause the production rate to plummet.

The production of herbaceous cotton as a function of the area planted in the state of Paraíba, Figure 1, shows that high production rates occur simultaneously with the use of considerable extensions in the cultivated area.

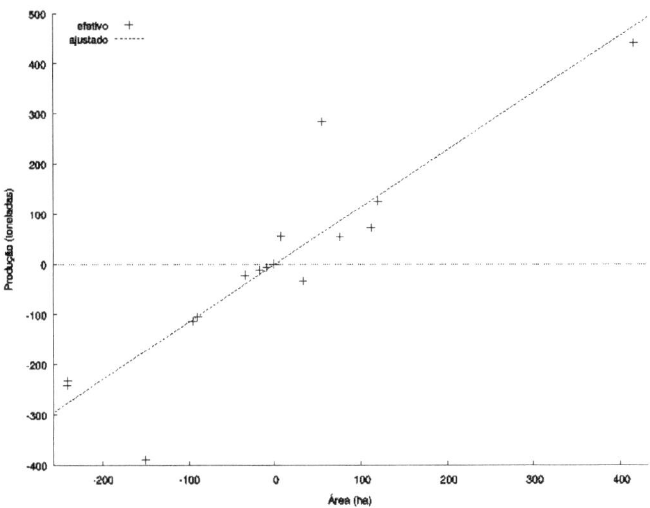

Figura 1. Production as a function of the effective and adjusted herbaceous cotton cultivated area in the state of Paraíba from January 2014 to December 2017.

Silva et al. (2005) report that in the state of Paraíba there are two favourable periods for preparing the soil and sowing herbaceous cotton: in January in the Sertão and in March in the Agreste region, but because of water restrictions sowing can be limited in some places.

Analysing actual and forecast production (Figure 2), it can be assumed that in some seasons expected production should have been higher than that obtained in that period.

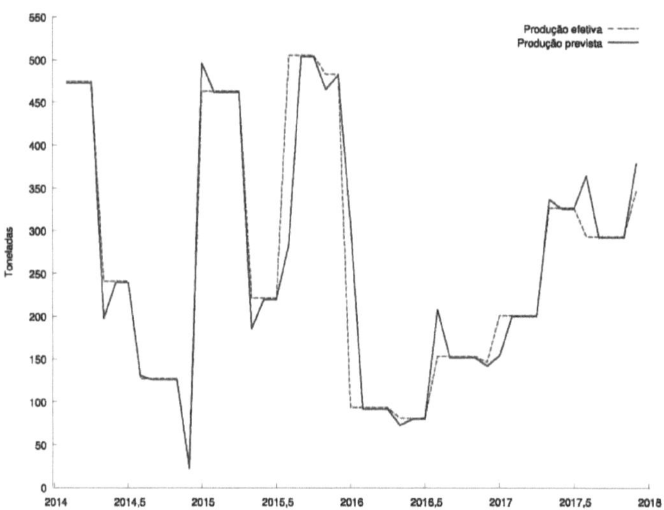

Figura 2.   Forecasting herbaceous cotton production in the state of Paraíba from January 2014 to December 2017.

- Peanuts

In the production of peanuts in the second crop in Paraíba as a function of planting area, it can be seen that 67.3 per cent (r-squared) justify the use of land extension in the production of this crop. The planting area was statistically significant at p<0.01 (Table 2).

Table 2. Ordinary Least Squares, with dependent variable peanut production in the second crop in the state of Paraíba from January 2014 to December 2017.

	Coefficient	p-value
Constant	4,14700	0,7364
Area (A)	1,77389	1,67e-012**
R-squared	0,673340	

**Significant at 5% ***Significant at 1%

Note: data subjected to first difference to become stationary.

Peanut production in the second crop has risen as the area under cultivation has increased (Figure 3), being directly proportional to the amount planted and harvested.

10

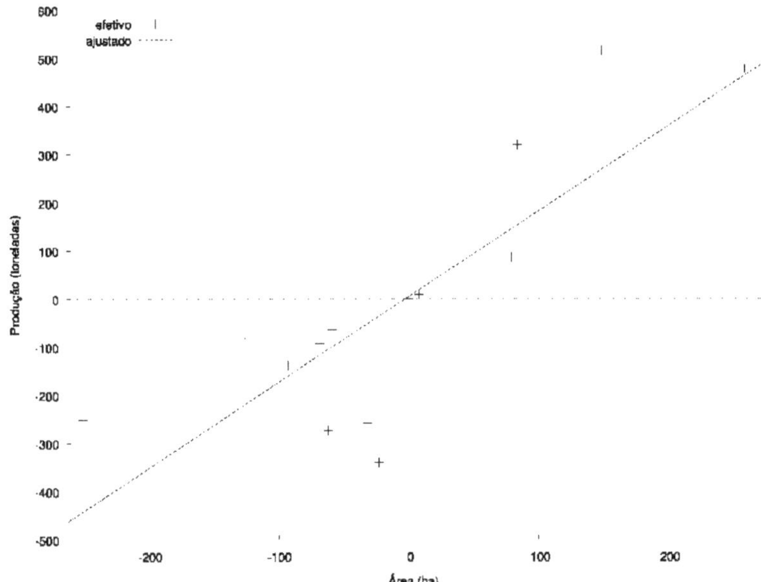

Figura 3. Production as a function of the effective and adjusted second crop peanut area in the state of Paraíba from January 2014 to December 2017.

The forecast for peanut production in Paraíba in the second harvest has some peaks and troughs, but in 2017 production tends to be lower than actual (Figure 4).

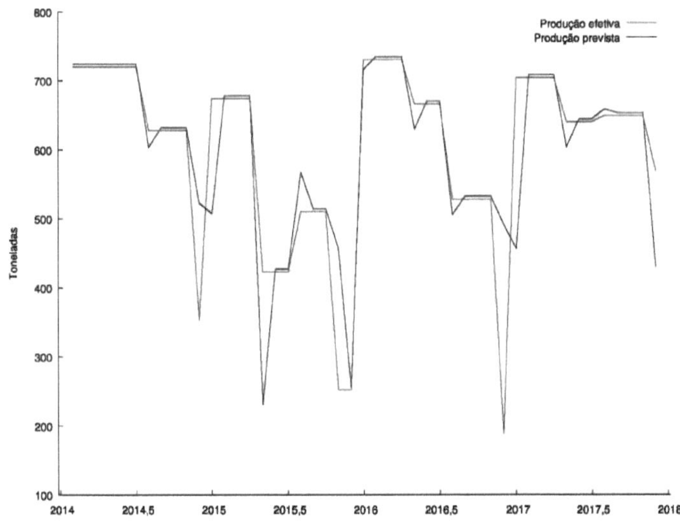

Figura 4. Forecasting peanut production in the state of Paraíba from January 2014 to December 2017.

According to CONAB (2017), the state of Paraíba planted more than 1,000 ha of peanuts in previous seasons, but water shortages have damaged several crops, reducing production and the area planted.

- Rice

Using Ordinary Least Squares (OLS), the area of rice cultivation in the state of Paraíba was statistically significant at 1%, with an r-squared of 95%, so this percentage shows that rice production justifies the amount of area applied to rice cultivation (Table 3).

Table 3 - Ordinary Least Squares, with the dependent variable peanut production in the second crop in the state of Paraíba from January 2014 to December 2017.

	Coefficient	p-value
Constant	-6,26617	0,9111
Area (A)	1,70984	2,01e-031***
R-squared	0,952524	

**Significant at 5% ***Significant at 1%

Note: data subjected to first difference to become stationary.

The area under rice cultivation is directly proportional to production (Figure 5), so an increase in production will mean an increase in the area planted. CONAB (2017) states that rice in the northeast can be grown in rainfed or irrigated conditions, which is an interesting alternative for regions with water deficits, but producers choose more profitable crops, which has reduced the area for rice cultivation and production, with planting beginning in December.

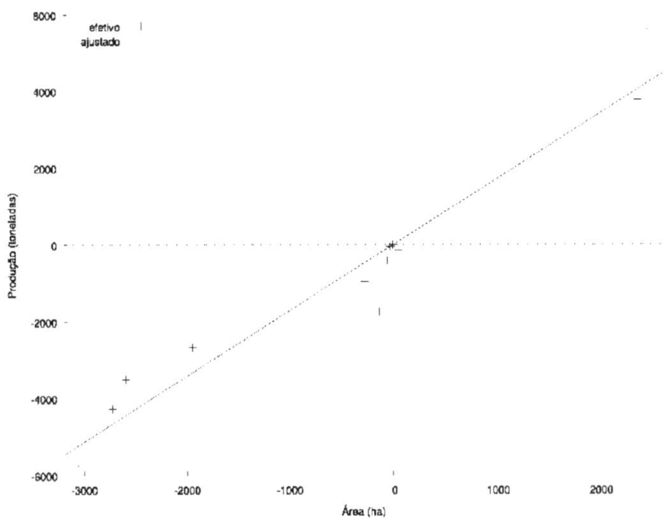

Figura 5. Production as a function of actual and adjusted rice acreage in the state of Paraíba from January 2014 to December 2017.

Irregular rainfall and the production of rice from other regions compete with that produced in the region, lowering costs and causing losses for farmers (CONAB, 2017).

In the rice production forecast for the period from January 2014 to December 2017 (Figure 6), it can be seen that the data obtained is similar, with only a few differences in scenarios that could have occurred. It can be seen that in the first half of the year there is a high production rate, since planting begins in December, according to CONAB (2017), so after harvesting the area can be reused for other short-cycle crops or prepared for the next planting.

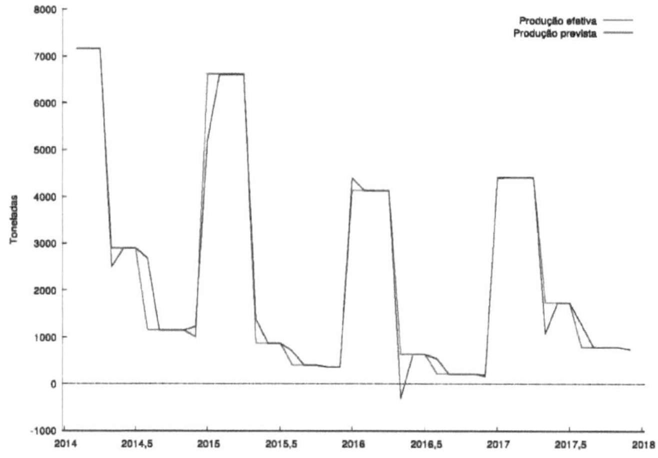

Figura 6.   Forecasting rice production in the state of Paraíba from January 2014 to December 2017.

- Beans, Iª and 2ª crop

Iª Bean harvest

In the first bean crop in Paraíba, R^2 is 76.4% (Table 4), so the dependent variable bean production and the explanatory variable area cultivated are clearly related and the remaining 23.6% is related to other variables.

Table 4 - Ordinary Least Squares, with the dependent variable bean production in the first harvest in the state of Paraíba from January 2014 to December 2017.

	Coefficient	p-value
Constant	-55,8241	0,9487
Area (A)	1,13990	1,00e-015**
R-squared	0,764418	

**Significant at 5% ***Significant at 1%

Note: data subjected to first difference to become stationary.

The number of beans produced in the first crop increases with the size of the area (Figure 7).

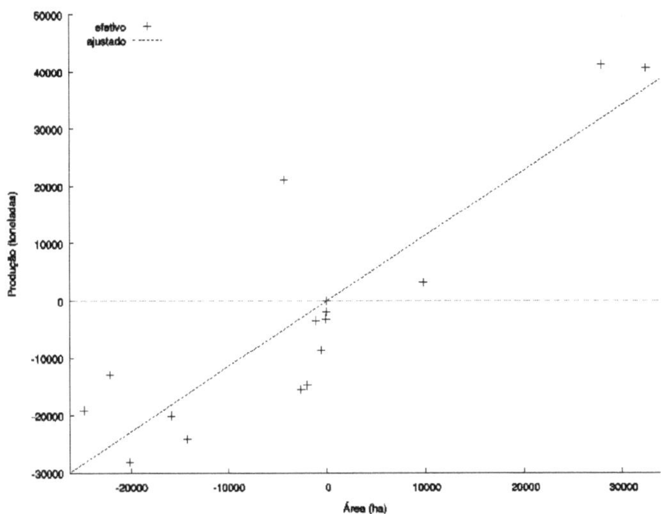

Figure 7. Bean production in the state of Paraíba from January 2014 to December 2017.

The time series from January 2014 to December 2017 with the forecast for this entire season, Figure 8, with the highest yields in the first half of each year, shows that these first months are favourable for bean production.

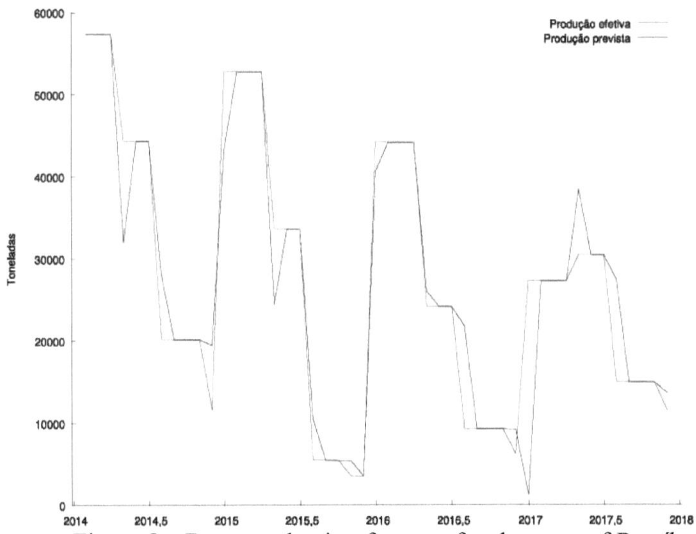

Figura 8. Bean production forecast for the state of Paraíba from January 2014 to December 2017.

Lima Filho et al. (2013) state that for the state of Bahia, the best time to sow cowpeas is between June and July, under the climatic conditions of this region. Rocha et al. (2017) report that cowpeas can have the longest cycle at 90 days, indicating that producers of large areas should choose species that ripen early.

2ª Bean harvest

In the second bean crop, the amount of bean production is related to the area cultivated by 73.16%, Table 5.

Table 5 - Ordinary Least Squares, with the dependent variable second crop bean production in the state of Paraíba from January 2014 to December 2017.

	Coefficient	p-value
Constant	55,8190	0,8927
Area (A)	0,998345	1,91e-014**
R-squared	0,731694	

**Significant at 5% ***Significant at 1%

Note: data subjected to first difference to become stationary.

Bean yields in the second crop are significant in relation to the area cultivated (Figure 9)

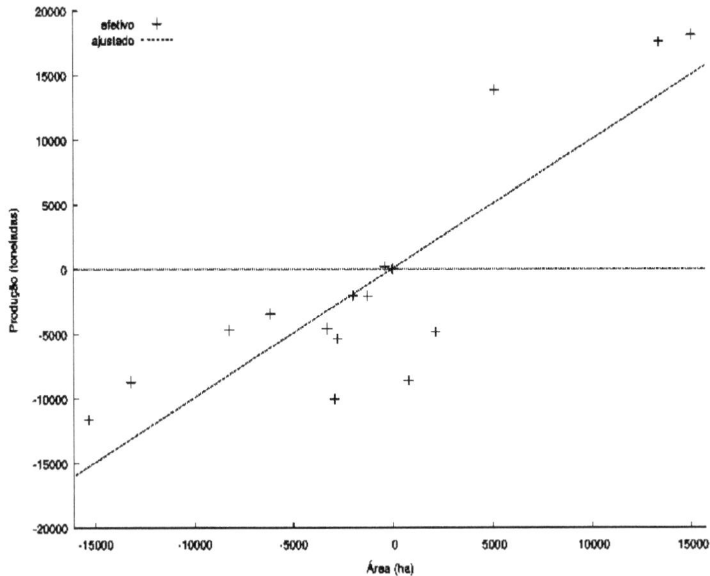

Figura 9. Second crop bean production in the state of Paraíba from January 2014 to December 2017.

Black beans are not widely grown in Paraíba and are affected by the lack of rainfall. In September 2017, according to CONAB (2017), there were 1.1 thousand black beans grown in the state.

hectares cultivated and already harvested. The caupi bean had a fairly extensive planted area, but because of the drought period, there was a reduction of around 63% of the area with almost all the production harvested or in the ripening phase in September, i.e. in the second harvest (CONAB, 2017).

It can be seen in Figure 10 that the forecast tended to have lower estimated quantities of beans than what was actually obtained.

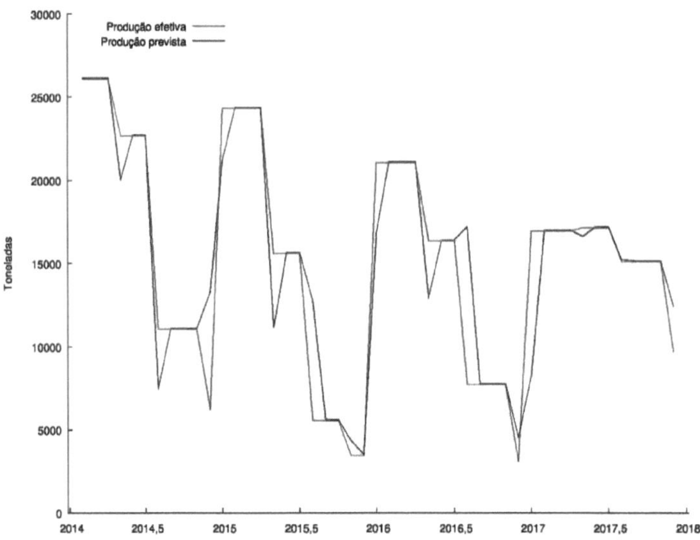

Figura 10.   Forecasting second crop bean production in the state of Paraíba from January 2014 to December 2017.

- Maize Iª harvest

The constant and the area were statistically significant at 1%, with R^2 of 70.3%, so the amount of maize production was obtained from almost the entire cultivated area, the losses were due to other factors, Table 6.

Table 6 - Ordinary Least Squares, with the dependent variable corn cultivated area in the state of Paraíba from January 2014 to December 2017.

	Coefficient	p-value
Constant	-122,901	0,9580
Area (A)	2,23548	1,95e-013**
R-squared	0,702755	

**Significant at 5% ***Significant at 1%

Note: data subjected to first difference to become stationary.

By increasing the area by 1 hectare, there is an increase in production of approximately 2 tonnes (Figure 11).

18

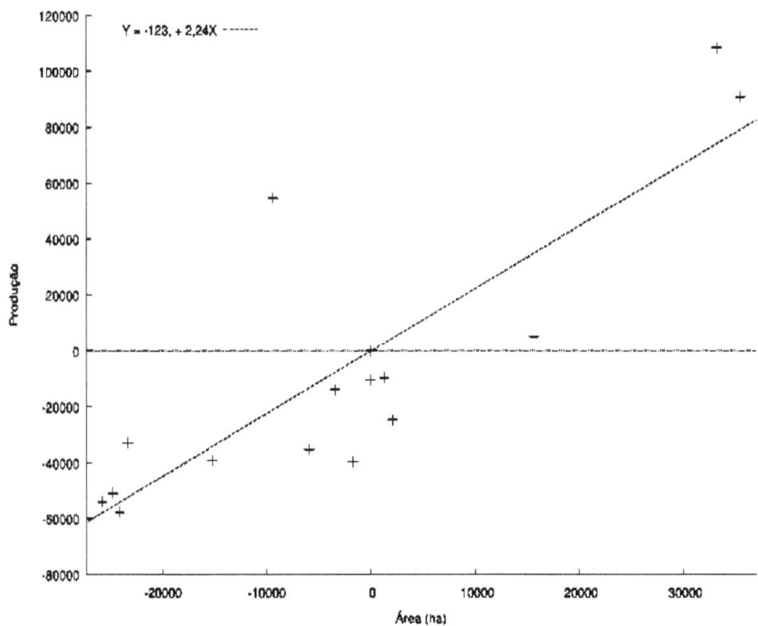

Figura 11.   Estimated maize production in relation to the area cultivated in the state of Paraíba.

In the production forecast for the months from January 2014 to December 2017 (Figure 12), it can be seen that there are few variations, with a sharp drop in production in the first half of 2017 and a rise in the second half of the same year.

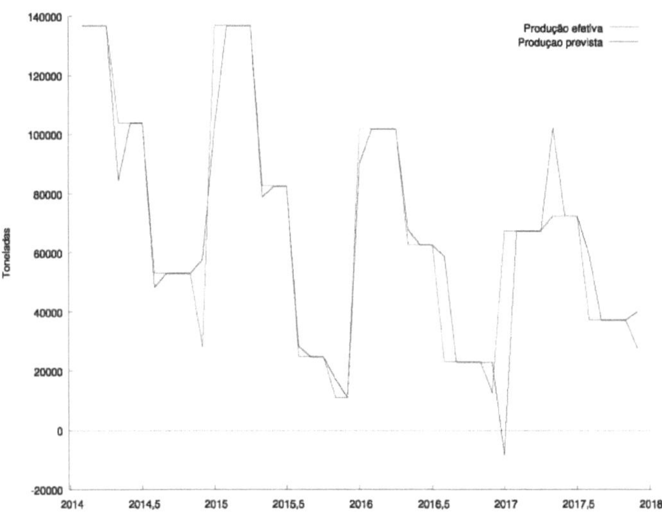

Figure 12. Maize production forecast in relation to the area cultivated in the state of Paraíba from January 2014 to December 2017.

- Banana

In the case of banana cultivation in Paraíba, Table 7 shows that there was no statistically significant effect, the OLS being within the established standards, such as the Ramsey test and heteroscedasticity, however banana production is not influenced by the area planted, other factors are involved in this crop.

Table 7. Ordinary Least Squares, with dependent variable banana cultivated area in the state of Paraíba from January 2014 to December 2017.

	Coefficient	p-value
Constant	-506,184	0,5759
Area (A)	2,03103	0,3114
R-squared	0,022765	

**Significant at 5% ***Significant at 1%

Note: data subjected to first difference to become stationary.

The area cultivated with bananas does not expand and production with the same

20

cultivated area fluctuates a lot (Figure 13), so it can be said that banana production is related to other parameters, where you can have high production without having to increase the area.

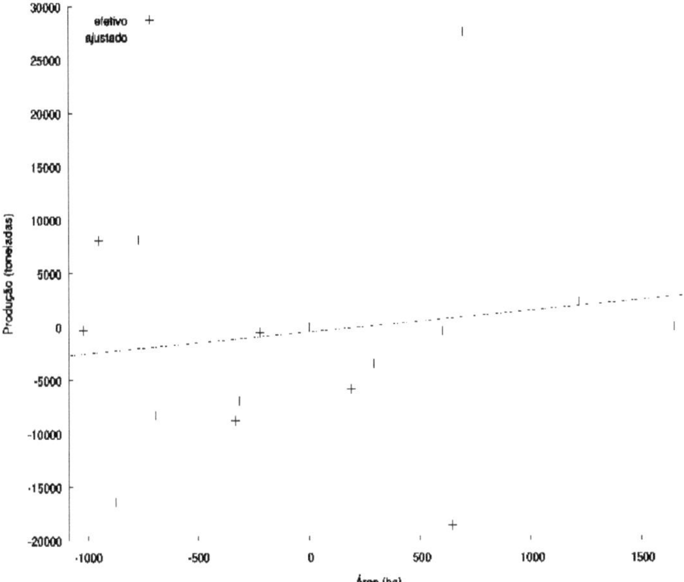

Figura 13.   Banana production in relation to the area cultivated in the state of Paraíba.

In the banana production forecast, Figure 14, it can be seen that in 2015 there was a sharp drop and the estimate was that it would be stronger, although there is a variation, a significant decrease is expected in the second half of 2017.

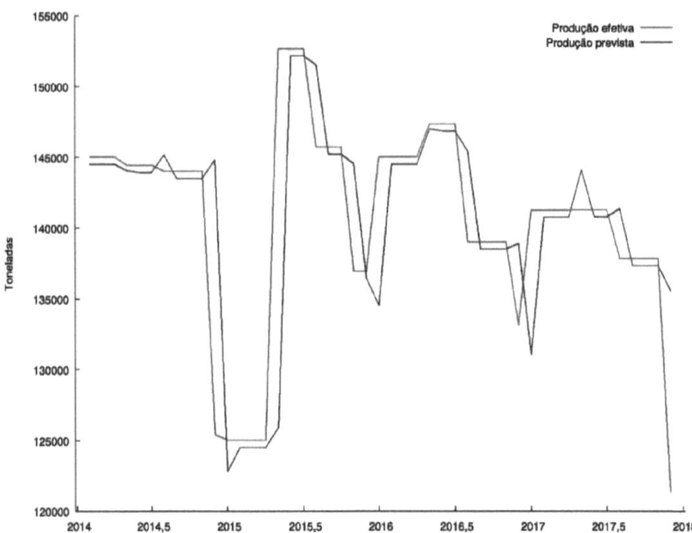

Figura 14. English potato production forecast in relation to the area cultivated in the state of Paraíba from January 2014 to December 2017.

The result of the correlogram of the residuals, Figure 15, shows no correlations, and the OLS model is considered to be well estimated, as was the case with the other agricultural productions analysed.

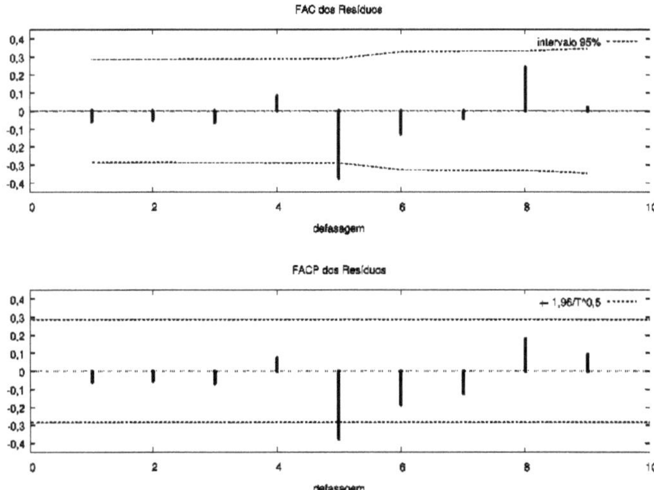

Figura 15. Correlogram of banana production residues as a function of cultivated area with a 95% confidence interval.

- English potato - 2ª harvest

22

With an R^2 of 93.7% and the explanatory variable statistically significant at 1% (Table 8), increasing the planting area increases production by 10.7 tonnes per hectare, making it one of the highest yielding crops per hectare.

Table 8. Ordinary Least Squares, with dependent variable area cultivated with English potatoes in the state of Paraíba from January 2014 to December 2017.

	Coefficient	p-value
Constant	11,0298	0,4433
Area (A)	10,7781	1,13e-028 ***
R-squared	0,937113	

**Significant at 5% ***Significant at 1%

Note: data subjected to first difference to become stationary.

Figure 16 shows increasing production in relation to the area cultivated, so if production is higher in a given area, productivity will be higher.

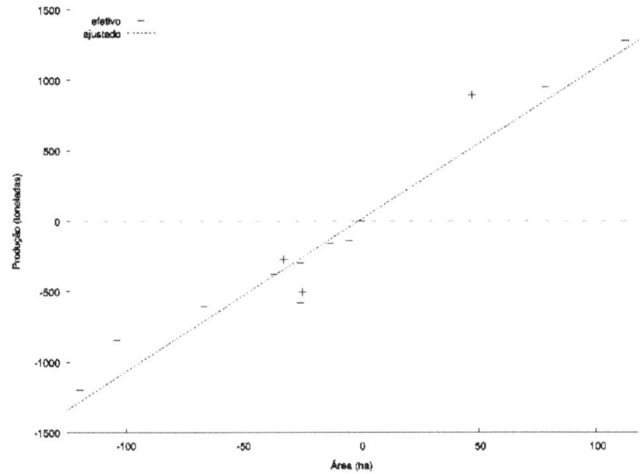

Figura 16. English potato production in relation to the area cultivated in the state of Paraíba.

The forecast for English potato production in the second harvest in Paraíba (Figure 17) shows few fluctuations, with lower production in the second half of 2017.

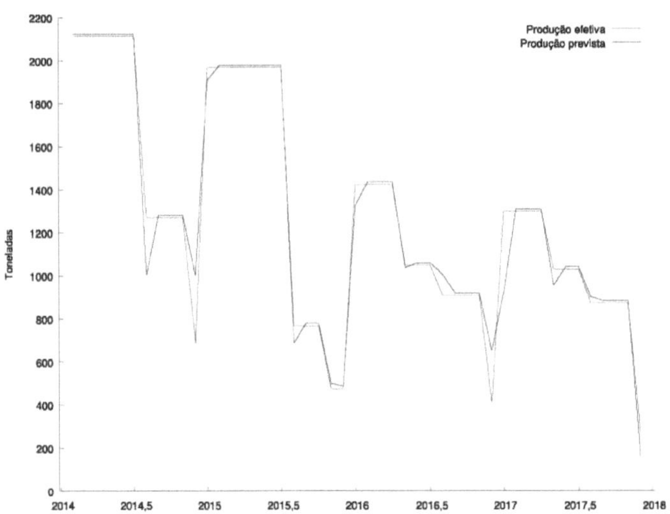

Figura 17. Forecast English potato production in relation to the area cultivated in the state of Paraíba from January 2014 to December 2017.

- Sugar cane

With no significant statistical effect, Table 9, the OLS test is adequate according to the tests applied. Although the area cultivated does not explain sugar cane production, it can be seen that 1 hectare increases production, although other factors interfere with this cultivar.

Table 9. Ordinary Least Squares, with the dependent variable area cultivated with sugar cane in the state of Paraíba from January 2014 to December 2017.

	Coefficient	p-value
Constant	-1,30475e+06	0,3189
Area (A)	3,99363	0,9812
R-squared	0,000013	

**Significant at 5% ***Significant at 1%

Note: data not subjected to first difference.

Sugarcane production rises as the area increases, but it can be seen in Figure 18 that the area is not the factor that influences production, but there are other variables that are negatively influencing it.

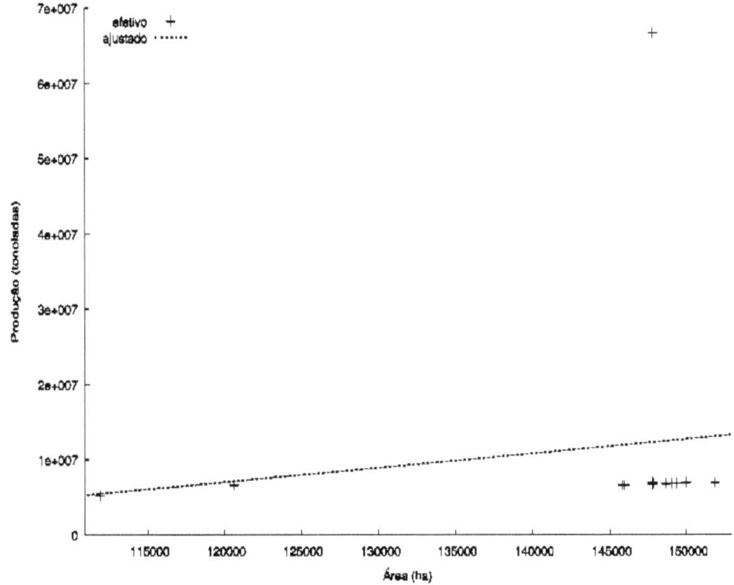

Figura 18.   Sugarcane production in relation to the area cultivated in the state of Paraíba.

The forecast for sugar cane production shows a drop in the second half of 2014 and then a stabilisation in production (Figure 19).

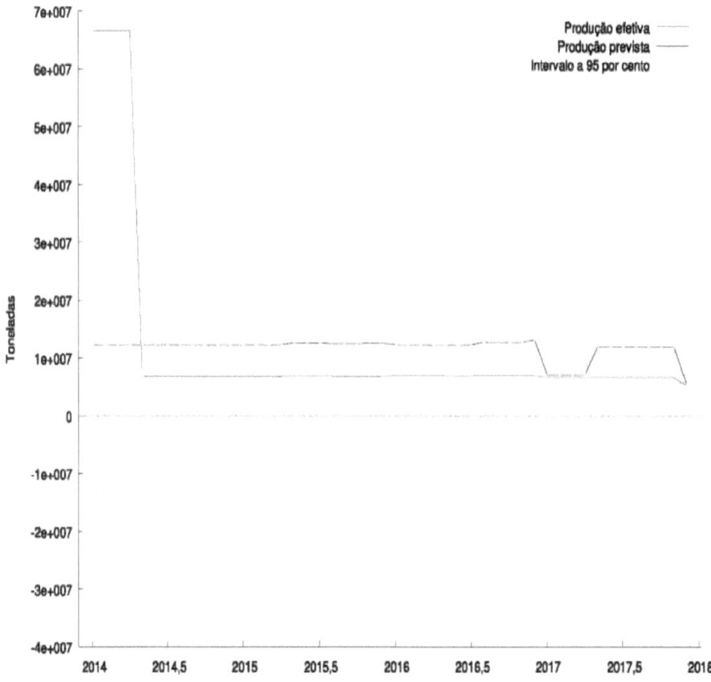

Figura 19. Sugarcane production forecast in relation to the area cultivated in the state of Paraíba from January 2014 to December 2017.

The residuals generated in the MQO analysis are within the 95% confidence interval. Thus the result of the correlogram of the residuals has no significant correlations other than zero (Figure 20).

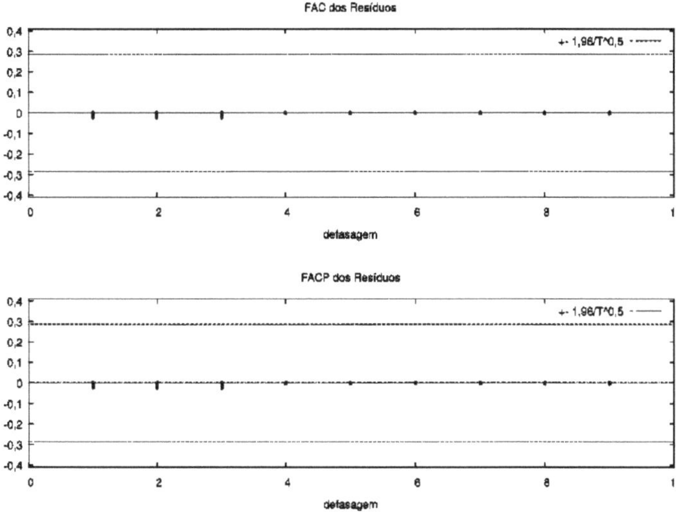

Figura 20. Correlogram of sugar cane production residues as a function of cultivated area with a 95% confidence interval.

- Cashew nuts

The area variable was statistically significant, Table 10. With an increase in plantation area there is an increase in cashew nut production, similar to what happens with other crops. The R^2 value was 52.3%, so the area explains more than 50% of the production obtained.

Table 10. Ordinary Least Squares, with dependent variable cashew nut cultivated area in the state of Paraíba from January 2014 to December 2017.

	Coefficient	p-value
Constant	1,76103	0,8247
Area (A)	0,526570	5,73e-09 ***
R-squared	0,533160	

**Significant at 5% ***Significant at 1%

Note: data not subjected to first difference.

As the area cultivated with cashew trees increases, so does the use of cashew nuts (Figure 21).

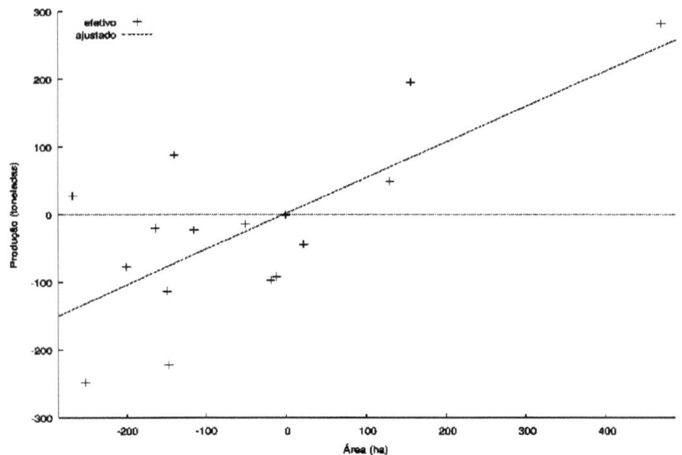

Figura 21. Cashew nut production in relation to the area cultivated in the state of Paraíba.

Estimated production was forecast to be lower in 2105 and 2017, but actual production fell the most in the second half of 2017 (Figure 22).

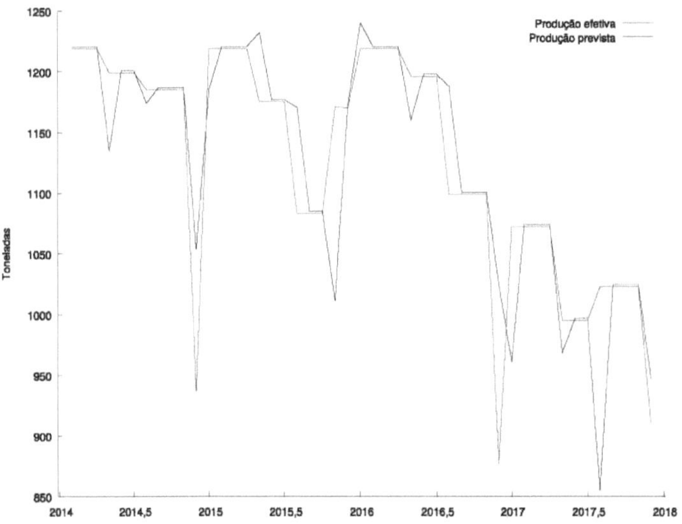

Figura 22. Cashew nut production forecast in relation to the area cultivated in the state of Paraíba from January 2014 to December 2017.

Figure 23 shows that the residuals generated in the OLS analysis are within the 95%

confidence interval. Thus, the result of the correlogram of the residuals has no significant correlations other than zero, and the bars are within the confidence interval, making the OLS a model considered to be well estimated, which is what happened with the other agricultural products analysed.

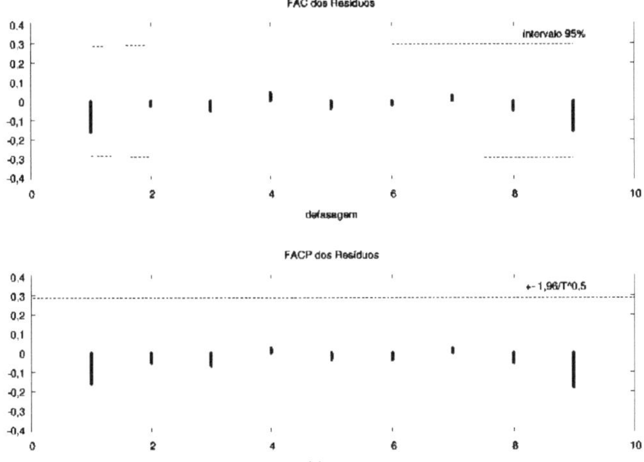

Figura 23. Correlogram of cashew nut production residues as a function of cultivated area with a 95% confidence interval.

Smoke

With 99% (R^2), the area justifies tobacco production in Paraíba (Table 11), with the area variable significant at 1%.

Table 11. Ordinary Least Squares, with dependent variable tobacco cultivated area in the state of Paraíba from January 2014 to December 2017.

	Coefficient	p-value
Constant	-0,193695	0,8585
Area (A)	0,976162	4,61e-048 ***
R-squared	0,991345	

**Significant at 5% ***Significant at 1% Note: data not subjected to first difference.

Tobacco production is growing as the area planted increases (Figure 24).

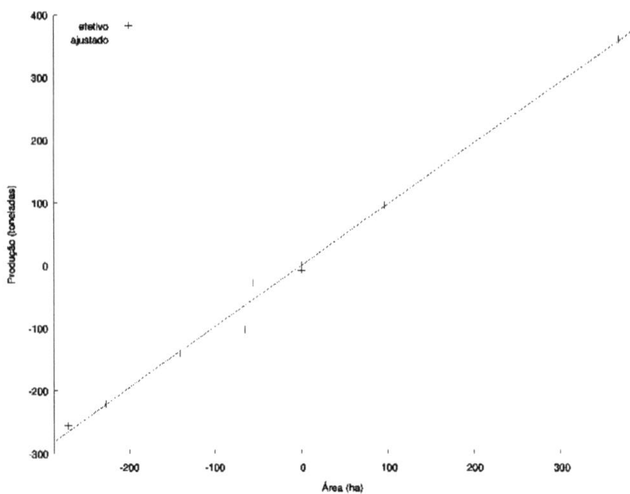

Figura 24. Tobacco production in relation to the area cultivated in the state of Paraíba.

The forecast shows that the first half of 2014 and 2015 saw the highest rates of tobacco production, with the lowest production recorded at the end of the second half of 2014, which remained stable from 2016 until the end of 2017 (Figure 25).

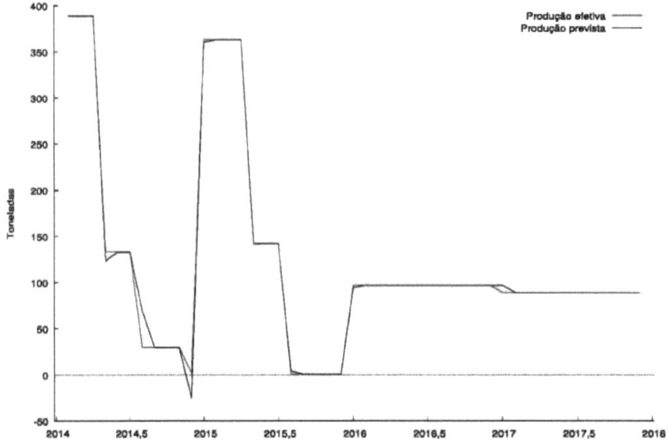

Figura 25. Tobacco production forecast in relation to the area cultivated in the state of Paraíba from January 2014 to December 2017.

30

Orange

Table 12 shows that the variables were statistically significant at the 1% level, with an R^2 of 69.6%, so around 70% of the total area used *is* responsible for this production, so raising 1 hectare increases the amount of oranges by approximately 14.7 tonnes.

Table 12. Ordinary Least Squares, with dependent variable orange cultivated area in the state of Paraíba from January 2014 to December 2017.

	Coefficient	p-value
Constant	-5879,51	1,27e-06 ***
Area (A)	14,7447	1,71e-013 ***
R-squared	0,696452	

**Significant at 5% ***Significant at 1%

Figure 26 shows that the highest yields were obtained in very extensively cultivated areas.

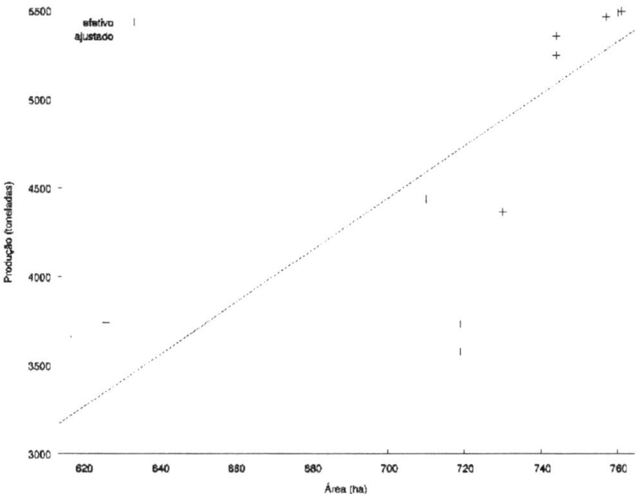

Figura 26. Tobacco production in relation to the area cultivated in the state of Paraíba.

Figure 27 shows that there was a sharp drop in orange production from the second half of 2016 onwards, with an increase in 2017, although with lower production figures when compared to the previous years analysed, which can be explained by water restrictions in this region.

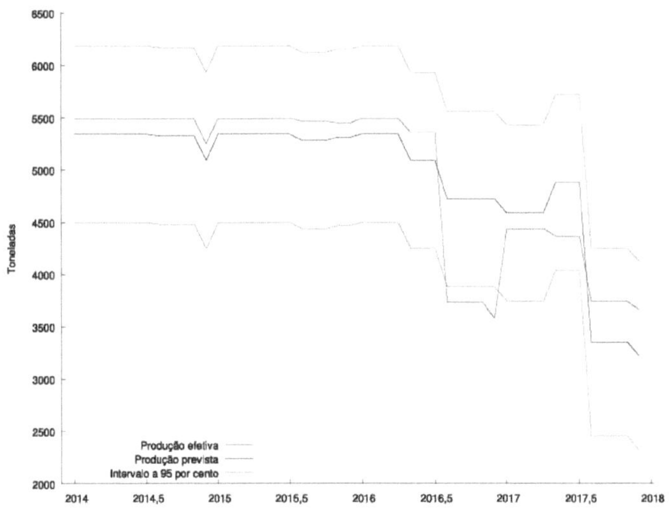

Figura 27.   Orange production forecast in relation to the area cultivated in the state of Paraíba from January 2014 to December 2017.

Cassava

When analysing cassava production as a function of cultivated area (Table 13), it can be seen that there was no statistically significant effect, with a very low r-squared, and as the cultivated area increases there is a reduction in cassava production, which may be related to water deficit, as adding area for planting directly increases the amount of water needed for irrigation.

Table 13. Ordinary Least Squares, with dependent variable cassava cultivated area in the state of Paraíba from January 2014 to December 2017.

	Coefficient	p-value
Constant	-11,9046	0,9893
Area (A)	-0,757110	0,2819
R-squared	0,025683	

**Significant at 5% ***Significant at 1%

Note: data not subjected to first difference.

In terms of cassava production, Figure 28 shows that there is a reduction in production depending on the area planted, so increasing the area directly affects the

32

production of this crop.

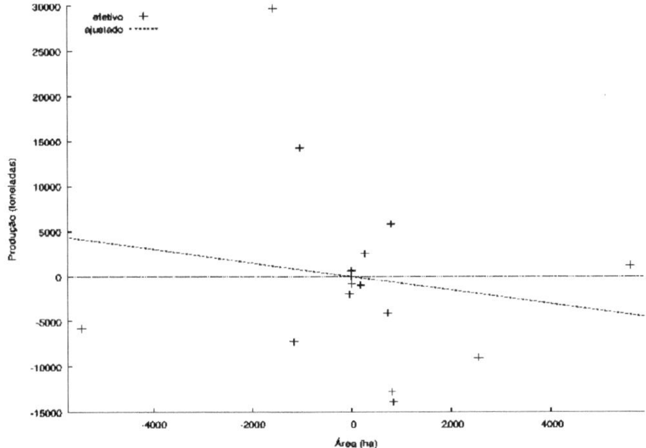

Figura 28.   Cassava production in relation to the area cultivated in the state of Paraíba.

When looking at the actual and adjusted cassava production, by ordinary least squares, Figure 29, the differences are few, with a greater decline at the end of 2015 and the beginning of 2016, and an increase in cassava production in the first half of 2016, the highest rate.

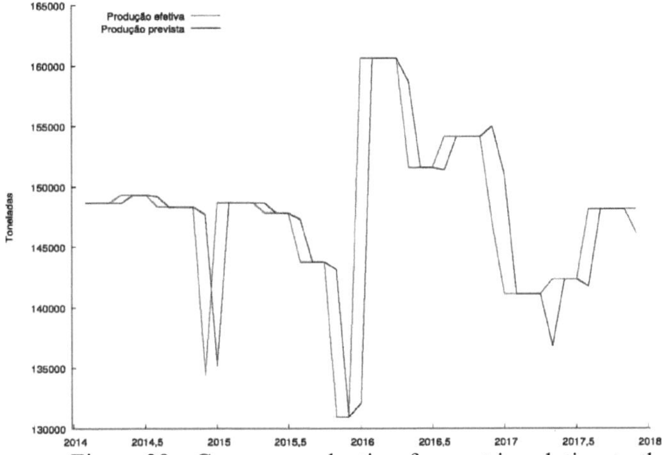

Figura 29.   Cassava production forecast in relation to the area cultivated in the state of Paraíba from January 2014 to December 2017.

Despite not having a statistically significant effect, in Figure 30, the residuals generated in the MQO analysis are within the 95% confidence interval, affirming that the MQO is an adequate model and is well estimated.

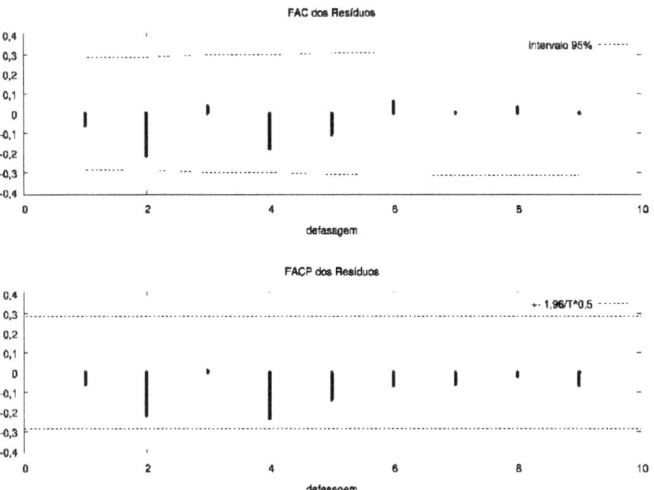

Figura 30. Correlogram of cassava production residues as a function of cultivated area with a 95% confidence interval.

Tomatoes

When checking tomato production as a function of the area used in Paraíba, Table 14 shows that the area variable is statistically significant (p<0.01). By increasing the planting area there is a likelihood of an increase in tomato production, with an R^2 of 91.25 per cent.

Table 14. Ordinary Least Squares, with dependent variable tomato cultivated area in the state of Paraíba from January 2014 to December 2017.

	Coefficient	p-value
Constant	-59,3018	0,3799
Area (A)	35,9701	1,90e-025 ***
R-squared	0,912573	

**Significant at 5% ***Significant at 1%

Note: data not subjected to first difference.

In Figure 31, tomato production as a function of area increases as the size of the

34

growing site increases.

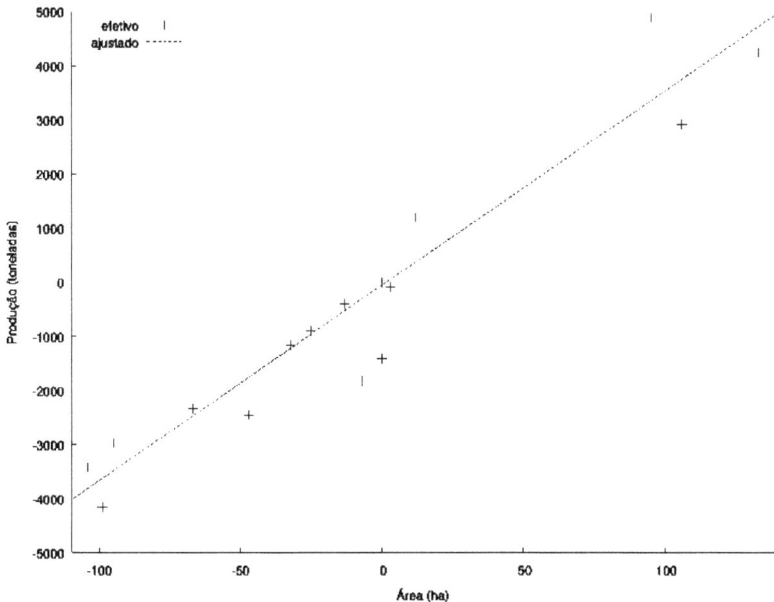

Figura 31. Tomato production in relation to the area cultivated in the state of Paraíba.

Expected tomato production in Paraíba was at its best in the first half of 2014, 2015 and 2016, with a sharp drop thereafter (Figure 32). s

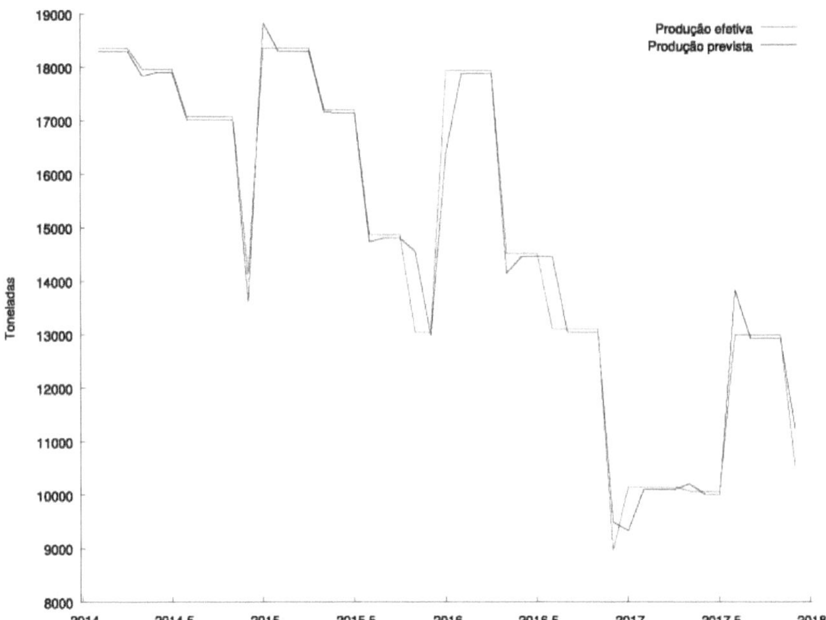

Figura 32.   Tomato production forecast in relation to the area cultivated in the state of Paraíba from January 2014 to December 2017.

CONCLUSION

The production of herbaceous cotton, peanuts, rice, beans, maize, bananas, potatoes, sugar cane, cashew nuts, tobacco, oranges, manioc and tomatoes all saw a reduction in productivity;

Tomatoes have the highest production due to the area planted, followed by oranges and English potatoes, so despite the water deficit they are cultivars that have seen a reduction in production, but with the highest yields;

By increasing the area under cultivation, there were high production rates, but other factors that influence the crops, such as rainfall, have a direct influence, since they depend on water for irrigation.

REFERENCES

CONAB - National Supply Company. Monitoring the Brazilian harvest. Fourth survey/January 2017, v.4, n.4, 2017,162p.

CONAB - National Supply Company. Monitoring the Brazilian harvest. Twelfth survey/September 2017, v.4, n.12, 2017, 162p.

LIMA FILHO, A. F.; COELHO FILHO, M.A.; HEINEMANN, A.B. Determination of sowing times for cowpea in the Recôncavo Baiano using the CROPGRO model. Brazilian Journal of Agricultural and Environmental Engineering, v.17, n.12, p.1294-1300, 2013.

GRETL. 2018.http://gretl.sourceforge.net/

OLIVO, A.M.; ISHIKI, H.M. Brazil in the face of water scarcity. Colloquium Humanarum, v. 11, n. 3, p.41-48, 2014. DOI: 10.5747/ch.2014.vll.n3.hl70

ROCHA, M.M.; SILVA, K.J.D.; FREIRE FILHO, F.R.; MENEZES JÚNIOR, J.A.N. Cultivars. Production system data, Embrapa, 2017, p.10.

SILVA, D. F.; ALCÂNTARA, C. R. Water Deficit in the Northeast Region: Spatial-Temporal Variability. Cient. Exatas Tecnol., v. 8, n. 1, p.45-51, 2009.

SILVA, M. T.; AMARAL, J. A. B. DO; BELTRÃO, N. E. DE M.; NASCIMENTO, M. G. do.(2005). Definition of the sowing time for herbaceous cotton (Gossypium Hirsutum L. R. Latifolium Hutch) in the state of Paraíba, according to climate risk zoning. EMBRAPA, Zoning of Agricultural Risks in Brazil, Agrometeorological Monitoring and Crop Forecasting.

Sustainable Development Indicators (SDI). Available at: https://sidra.ibge.gov.br/pesquisa/ids/documentos .2017.

VENANCIO, D.F.V.; SANTOS, R.M.; CASSARO, S.; PIERRÔ, P.C.C. The water crisis and its global contextualisation. Enciclopédia Biosfera, v.ll, n.22, p.1-13, 2015.

CHAPTER 2

LIVESTOCK WATER REQUIREMENTS IN THE SEMIARID REGION OF PARAIBANO -PB

Thalis Leandro Bezerra de Lima, Júlia Soares Pereira, Lilian de Queiroz Firmino, Dihego de Souza Pessoa, Viviane Farias Silva & Vera Lucia Antunes de Lima

SUMMARY

The municipality of Fagundes is located in the state of Paraíba, Brazil, in the geoenvironmental unit of the Borborema Plateau, with altitudes ranging from 500 to 800 m, an area of 162 km^2 and 11,830 inhabitants. It has the climate of the caatinga biome, which is called semi-arid. The characteristics of this type of climate are low humidity and low rainfall. This irregular climate influences the course of rivers, which dry up at certain times, reduces the availability of water for plants, animals and humans, and increases the aridity of the environment. The climate is therefore a determining factor in the caatinga: it ends up defining the landscape and the habits of the inhabitants of this biome. In this context, the aim of this study was to analyse the water demand of livestock farming in the semi-arid region of Paraíba, with a focus on the municipality in question. Knowing the water demand of the population and livestock improves the management of water resources and the abstraction of water for these purposes, as well as a way of planning for the future. It was observed that in the municipality of Fagundes-PB, poultry is suitable for rearing because it has a lower water demand and greater profitability, while cattle are not recommended due to the high water demand required for their maintenance since, according to the record of recent years of drought and prolonged drought, there is a water deficiency in the municipality.

Keywords: Water deficit; Animal vulnerability; Drought; Caatinga biome.

INTRODUCTION

Drought is a natural phenomenon that occurs quite frequently in the semi-arid region of the Northeast. One of the activities that influenced the settlement of this area was livestock farming. It is assumed that the cattle herd is approximately 14 million head, with half of the herd in the Northeast, representing 15% of the national herd, while in relation to goat and sheep herds the region is home to approximately 12 million animals, considering these animals to be better adapted to the semi-arid conditions of the region (IBGE, 2006). The water deficit causes various economic, agricultural and social problems and is a factor that limits the number of animals in the region.

The Caatinga is the predominant vegetation in the semi-arid region, with a high genetic diversity of animals and plants, most of which are endemic species. Araújo Filho (2006) reports that animal feed is mainly grazed by forage plants and can last for a short period of time. Santos et al. (2011) cite the following forage plants: sorghum, leucine, gliricidia, guandu, algaroba, current grass, gramão grass, bufei grass and forage palm. These species are considered exotic because they are not native to the region.

The semi-arid region is characterised by scarce and irregular rainfall, shallow and stony soil and a semi-arid climate. These edaphoclimatic conditions, with high temperatures and direct solar radiation over a long period of the year, cause stress in the animals, affecting production, according to Guerrini (1981), and Souza et al. (2005) state that in order to obtain favourable production, the animals need to have adequate thermal comfort conditions. Coexistence with drought applies technologies and alternatives so that people living in these regions have adequate living conditions and the possibility of producing and earning an income. Holanda Júnior et al. (2004) report that when food is produced, income and jobs are generated, people remain in rural areas, reducing the rural exodus, as well as offering quality of life. Knowing the water demand of the population and animals helps in the management of water resources, water abstraction for these purposes, as well as a way of planning for the future.

In semi-arid regions, the vulnerability of populations to climate variations is a serious problem. As such, the uncertainties generated by global warming reinforce the urgency of seeking ways to deal with current climate variability by strengthening resilience and reducing vulnerability (RIBOT et al, 2006).

Living with drought requires being prepared for long droughts and low-cost technologies, such as cisterns and water reuse, so that people have a decent quality of life to continue living in these regions. According to Anjos (2013), some limiting factors interfere with the acquisition of social technologies by those most in need. Souza et al. (2016) report that there are several technologies that can be used by the population in rural and urban areas, including underground dams, cisterns, septic tanks and others. For water collection and storage, cisterns and underground dams, according to Souza et al. (2016) are the ones that stand out, since water is the limiting factor in the semi-arid region for its socio-economic development.

In this context, the work was carried out with the aim of analysing the water demand of livestock farming in the semi-arid region of Paraíba, with a specific case for the town of Fagundes-PB.

MATERIAL AND METHODS

The study was carried out in the municipality of Fagundes, in the state of Paraíba, which, according to CPRM (2005), is part of the Borborema Plateau geoenvironmental unit, with altitudes ranging from 500 to 800 m, with an area of 162 km^2 , located at the following geographical coordinates: 7° 20' 45.56"S, 35° 47' 51.13"W (NOBRREGA et al., 2011). According to IBGE (2010), the municipality has approximately 11,830 inhabitants (Figure 1).

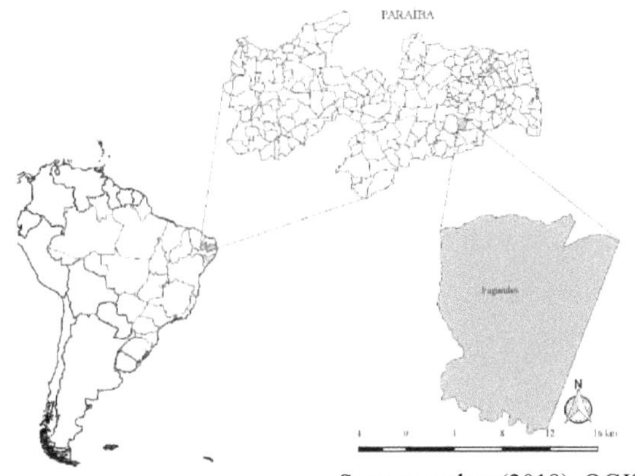

Source: author (2018), QGIS software.

Figura 1. Geographical location of the municipality of Fagundes-PB.

The information was obtained from the 2016 municipal livestock survey database obtained by IBGE (2016), as well as the average consumption in litres of each animal obtained by EMBRAPA (2013), the water demand of each animal, total and estimated water abstraction in the territorial area of the municipality of Fagundes-PB were estimated.

The main body of accumulated water is the Gavião reservoir in the middle reaches of the Paraíba River, with a maximum capacity of 1,450,840 m^3 , and currently a volume of 398,274 m^3 (27.45%), recorded on 21/07/2018 (AESA, 2018).

RESULTS AND DISCUSSION

Table 1 shows six types of livestock in the municipality of Fagundes, with poultry, at 40,000 head, having the lowest demand for water.

Table 1. Production animals in the municipality of Fagundes.

ANIMALS OF PRODUCTION	NUMBER OF HEADS	DEMAND/CONSUMPTION LITRES/DAY
POULTRY	40000	2
CATTLE	5400	45
GOATS	900	7
EQUINES	400	30
SHEEP	1700	7
PIGS	2500	7

Cattle are the largest animals, the second largest type of animal raised and with the highest demand for water, which limits their breeding and in times of drought they suffer the most and cause the greatest losses for the producer.

Silva et al. (2014) state that in times of water scarcity, farmers lose all or part of their agricultural production, as well as weakening their herds due to lack of food and little water, and often most of the animals die. These authors report that one of the ways of managing to live in regions with these characteristics is to plan and use drought coexistence techniques.

Approximately 900 goats were found in the region with an average consumption of 7 litres/day. Carneiro et al. (2016) report that the breeding of these animals in the semi-arid region is growing, which is explained by their adaptation to the region's soil and climate conditions, together with the farmers' management. Similar results were obtained by Aquino et al. (2016) in relation to the number of goats being lower than sheep in the semi-arid region. These authors observed that the number of cattle was lower than that found in this study, confirming that medium-sized animals predominate in this region.

Figure 2 shows that the daily water consumption of the more than 5,000 cattle is higher, at around 243 m^3 /day, although they consume less water than the poultry. The 40,000 birds consume approximately 80 m^3 /day. Poultry has a lower water demand despite being the majority of animals, with easy management and a shorter conversion time into income compared to other animals. Poultry is one of the animal species recommended for producers in regions with water deficits.

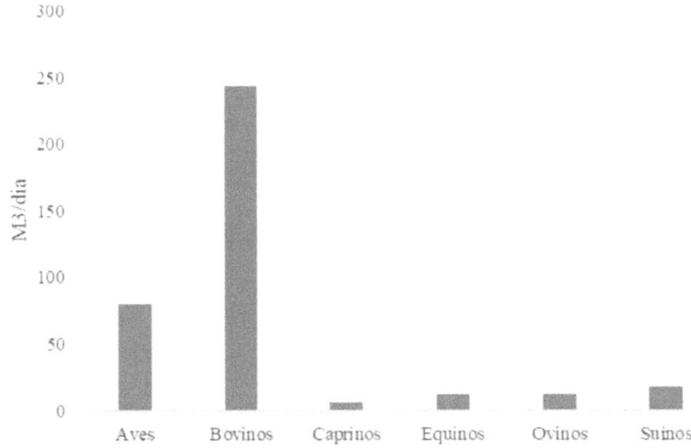

Figura 2. Daily water demand of production animals in the municipality of Fagundes - PB.

Silva et al. (2017), when researching water demand in the municipality of Boqueirão, reported that chickens are more common and cattle have a higher water demand.

The total annual water demand required for livestock production (Figure 3) shows that, in order to raise livestock properly without water stress, which would result in production losses and even death of the animals, it is essential to have around 135,000 m^3 /year to meet the water demand of the region's animals, but only 13,000 m^3 /year are available during times of maximum rainfall.

It can be seen that the amount of water available in the region makes it difficult to meet the water demand of livestock farms, making it necessary to collect water from another region, which increases the costs of raising animals.

Similar results were obtained by Silva et al. (2017) when analysing the water demand for livestock in the municipality of Boqueirão, where they found that the water available will only supply around 7% of the amount of water required by livestock, thus creating a huge deficit.

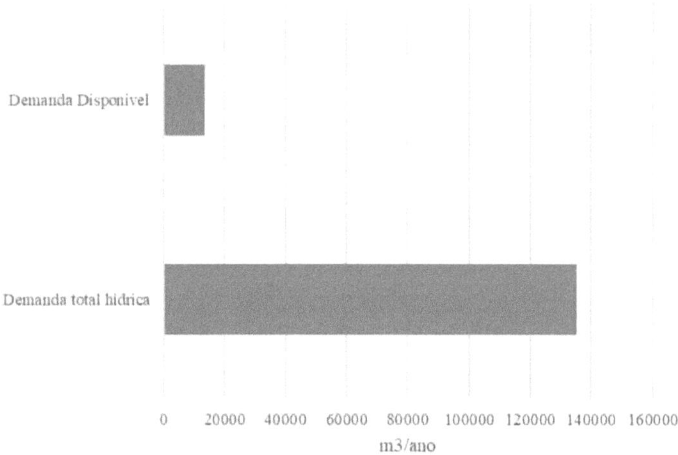

Figura 3. Total water demand required by livestock production in the municipality of Fagundes - PB.

Farias et al. (2017) state that there was a decrease in the number of animals in the drought period (2012 and 2013), with the decrease in water and food in this period, the animals were sold at lower market values and the others who tried to keep their herds had losses with the death of the animal, from thirst and/or hunger. That's why it's important to check which animals are suitable for breeding, especially in times of drought, in the most critical years.

CONCLUSION

Poultry is suitable for breeding because it has a lower water demand and higher profitability, while cattle are not recommended due to the high water demand required for their maintenance;

There is a water deficit in livestock production in the city of Fagundes.

BIBLIOGRAPHICAL REFERENCES

AESA. Executive Water Management Agency of the State of Paraíba. *Monitoring the Latest Volumes.* Available at: http://www.aesa.pb.gov.br/aesa-website/monitoramento/ultimos-volumes/. Accessed on 27 July 2018.

ANJOS, S. *Tecnologias e Projetos para Conviver com o Semiárido.* Available at: < http://www.insa.gov.br/wp-content/uploads/2013/05/0- Povo-II.pdf >. Accessed on: Jun.2018.

AQUINO, R.S.; LEMOS, C.G.; ALENCAR, C.A.; SILVA, E.G.; LIMA, R.S.; GOMES, J.A.F.; SILVA, A.F. *The reality of goat and sheep farming in the Brazilian semi-arid region: a portrait of the Araripe hinterland, Pernambuco. Publications in Veterinary Medicine and Zootechnics,* v.10, n.4, p.271-281, 2016.

ARAÚJO FILHO, J. A. (2006). *The Caatinga Biome.* In: Falcão Sobrinho, J. F.; Falcão, C. L. C. (Ed.). *Diversities, weaknesses and potential.* Sobral: Sobral Gráfica, p. 49 - 70.

CARNEIRO, W.P.; RAMOS, J.P.F.; PIMENTA FILHO, E.C.; CARVALHO, J.E.C.; MOURA, J.F.P. *Productive and reproductive evaluation of dairy goats in the semi-arid region of Paraíba. Revista Científica de Produção Animal,* v.18, n.l, p.18-25, 2016.

CPRM. Geological Survey of Brazil. Registration of groundwater supply sources project. *Diagnosis of the municipality of Lagoa*

Drought, state of Paraíba. Recife: CPRM/PRODEM, 2005. Available at: <http://www.cprm.gov.br>. Accessed on: 25 July 2015.

EMBRAPA. *Water consumption in animal production.* 2013, 6p.

FARIAS, A.A.; SOUSA, F.A.S.; MORAES NETO, J.M.; ALVES, A.S. *Droughts and their impacts in the municipality of Boqueirão, PB, Brazil. Revista Ambiente &Água,* v.12, n.2, p. 316-330, 2017.

GUERRINI, V. H. *Food intake of sheep exposed to hothumid, hot- dry and cool-humid environments.* American Journal of Veterinary Research, v.42, n.4, p.658-661, 1981.

HOLANDA JÚNIOR, E. V.; OLIVEIRA, C. A. A. V.; SILVA, P. C. G.; GUEDES, C. T. S.; ARAÚJO, G. G. L.; SILVA, C. N.; CEZIMBRA, C. M. (2004). *Typology and income structure of family-based goat and sheep farmers in the São*

Francisco hinterland of Bahia. In: Encontro da Sociedade Brasileira de Sistemas de Produção, 6, Anais... Aracaju: Embrapa Tabuleiros Costeriso. CD-ROM.

IBGE (2006). Brazilian Institute of Geography and Statistics. *Municipal livestock production.* Rio de Janeiro.

IBGE. *Brazilian Institute of Geography and Statistics. Demographic Census.* Livestock 2016. Available at: http://cidades.ibge.gov.br/xtras/perfil.php?lang=&codmun=250250&search =paraiba.|Boqueirao Accessed on: 27/08/2017.

RIBOT, J.C.; NAJAM, A.; WATSON, G. Climate variation, vulnerability and sustainable development in the semi-arid tropics. Cambridge University Press, Cambridge, United Kingdom and New York, 2006, p. 13-54.

SILVA, J.G.S.; PAULA, L.A.M.; ESMERALDO, G.G.; MONTE, F.C.D. Impacts of drought on settled family production units. Revista de desenvolvimento económico, v.16, n.30, 2014.

SILVA, V.F.; BEZERRA, C.V.C.; LIMA, T.L.B.; LIMA, R.W.C.; LIMA, V.L.A. *Analysis of the temporal variability of rainfall with the water demand of northeastern livestock.* In: III International Workshop on Water in the Brazilian Semi-Arid, Campina Grande-PB, 2017.

SOUZA, E. D.; SOUZA, B. B.; SOUZA, W. H. *Determination of physiological parameters and thermal gradient of different genetic groups of goats in the Semi-Arid Region.* Ciência e Agrotecnologia, v.29, n.l, p. 177-184. 2005.

SOUZA, N.G.M.; SILVA, J.A.; MAIA, J.M.; SILVA, J.B.; NUNES JÚNIOR, E.S.; MENESES, C.H.S. Social technologies for the development of the Brazilian semi-arid region. Journal of Biology & Pharmacy and Agricultural Management, v.12, n.3, p.1-12, 2016.

CHAPTER 3

RAINFALL PROGNOSIS IN SEMI-ARID PARAIBA

47

Carlos Vaillan Castro de Bezerra, Thalis Leonardo Bezerra de Lima, Dihego Souza Pessoa, Aldair dos Santos Gomes, Viviane Farias Silva and Vera Lucia Antunes de Lima

SUMMARY

Rainfall in the Brazilian semi-arid region is marked by spatio-temporal variation, which, combined with the low annual totals over the region, results in the frequent occurrence of long periods of drought, which can influence the permanence of the population in the area or the development of economic activities. The aim of this study was to analyse the monthly rainfall series to forecast precipitation for the months of May to October 2018 using an autoregressive model in the municipality of Camalaú, located in the Microregion of Cariri-Occidental and in the Mesoregion of Borborema, approximately 330 km from the capital, João Pessoa, in the state of Paraíba. It has a semi-arid climate, hot and dry with scarce and irregular rainfall throughout the year. The model adopted is widely used in studies of climatological variables because they show temporal and spatial correlation and can identify catastrophic events or periods of atypical weather in advance, minimising risks. Using the Gretl 2018a software, it was identified that in the months of June, July and October there is a possibility that there will be no precipitation records and that for the months of May, June and July the precipitation rates may be below 20 mm.

Keywords: climatological variables, dry and rainy periods, climate forecasting.

INTRODUCTION

Of all the meteorological phenomena that occur on the earth's surface, rainfall is the most important way for humanity to obtain water, because the rainfall regime defines the possibility and success of growing vegetables, fruit trees and other economic activities carried out to feed the population. It is also a limiting agent for the permanence of the population in a given area or even the installation of industrial activity. As defined by Silva (2005), precipitation *is* the water that falls from the atmosphere onto the Earth's surface, whether in liquid or solid form.

Completing the hydrological cycle, precipitation is the phase in which water returns once it has evaporated from the earth's surface and the oceans, condenses in the form of clouds and at a certain point falls back to the surface. Moura et al. (2007) mention that relative humidity and atmospheric pressure are parameters that together define the occurrence and intensity of precipitation phenomena, where an increase in one variable and a decrease in the other leads to rainfall. In this sense, precipitation is a meteorological variable that is complex to define and equally important to know. Although the way to measure the amount of rainfall is relatively simple using instruments known as rain gauges, from the oldest to the most modern, the way the data is collected is difficult to observe because the data needs to be representative for the area under study, so that for a city or region, for example, there are too many samples to say how much rainwater actually fell at each point, and errors in the installation and display of the instruments in the meteorological stations are also points of error in this observation (OLIVEIRA et al., 2014).

A set of rainfall observations in a given region organised in equal time intervals is known as a time series. This observation is important for the region under study because it enables knowledge and planning for local meteorological events, and due to the great difficulty of collecting data and studying forecasts, the study of such time series in order to analyse and model the rainfall series leads to the continuous development of representative prediction models, since their future behaviour cannot be predicted exactly but only estimated (SILVA, GUIMARÃES & TAVARES, 2008). Analysing time series of rainfall data is of great importance to people living and working in the area, as the occurrence of catastrophic events or periods of atypical weather affect the lives of the population, agricultural production and planning, as well as the establishment of companies in the industrial sector, which prefer areas with a more stable supply of rainfall and less risk of natural disasters (LOPES & SILVA, 2013; SILVA, GUIMARÃES & TAVARES, 2008).

In the state of Paraíba, as in the rest of the state, there is spatial variation in the occurrence of rainfall, which often leads to the effect of isolated rains in certain areas, and this is caused by the different atmospheric systems that act on the east coast of the Brazilian Northeast bathed by the Atlantic Ocean (MENEZES et al., 2008). In their study of these atmospheric systems, Roucou et al. (1996) observed that the occurrence of rainfall in the Brazilian Northeast is mainly affected by air movements along the equator and by the Intertropical Convergence Zone. In addition, according to the authors, the annual and longer-term distribution of rainfall in the micro-regions of Paraíba is similar to that which occurs in other states, with the rainy seasons being defined and expected in certain months of the year. For example, rain is expected from January to March in the Sertão, while in Cariri and Agreste the rainy season is from April to June.

The aim of this study was to analyse the monthly rainfall series to forecast precipitation using an autoregressive model for the municipality of Camalaú in the semi-arid region of Paraíba.

MATERIAL AND METHODS

The site chosen for the research was the municipality of Camalaú, located in the Microregion of Cariri-Occidental and in the Mesoregion of Borborema in the state of Paraíba. It is characterised by gentle reliefs, according to Patriota (2014), a semi-arid tropical climate and hyperxerophilous caatinga vegetation and parts of deciduous forest. The municipality of Camalaú, as stated by Galvinício et al. (2008), is part of the Alto Paraíba sub-basin.

The monthly rainfall information recorded in the municipality of Camalaú was obtained from the database of the Paraíba State Executive Water Management Agency (AESA), from January 2010 to April 2018. The data was processed for rainfall forecasting using Gretl 2018a software, with the autoregressive moving average model -ARMA adopted (2,2) or Auto-regressive Integrated Moving Averages - ARIMA (2,0,2). The forecast was made for the month of May 2018 to October 2018.

RESULTS AND DISCUSSION

Figure 1 shows the actual and adjusted rainfall data, with the highest rainfall in April 2014 at 223.30 mm, out of 100 months of recorded data.

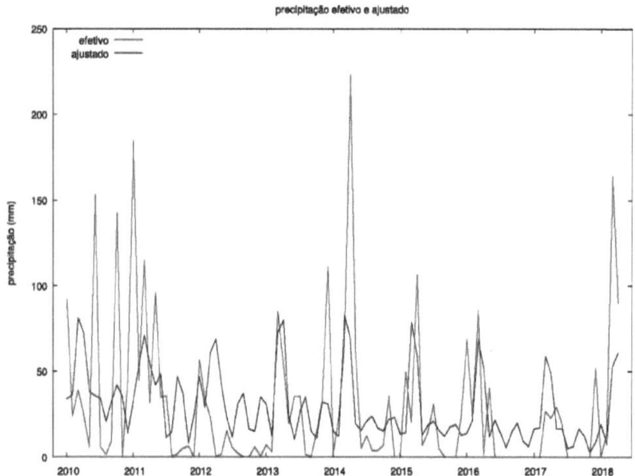

Figura 1. Actual and adjusted monthly rainfall from January 2010 to April 2018.

Bezerra et al. (2017) analysed the probability of maximum and minimum rainfall in the municipality of Camalaú/PB and found that the probability of maximum and minimum rainfall was 28.57%. Checking the rainfall variability of the municipality of Boqueirão in the same state over a 15-year period, Silva et al. (2016) found that the probability of a 600 mm rainfall is 13.3%.

From the rainfall data obtained, future rainfall data is based on previously recorded rainfall, so Figure 2 shows that rain is forecast for the months of May (19.6 mm), August (19.7 mm) and September (14.2 mm) in 2018, but in the months of June, July and October there may be no rainfall or rainfall within the 95% confidence interval.

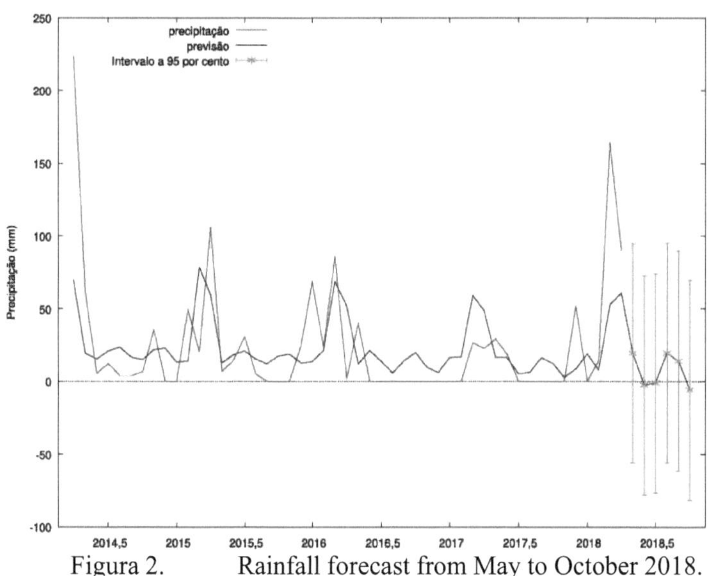

Figura 2.   Rainfall forecast from May to October 2018.

According to Bezerra et al. (2017), the chance of precipitation in September in this municipality is 57%. These authors also state that the chances of below-average rainfall are 100 per cent.

So for farmers, it's a way of getting an overview of rainfall possibilities for agricultural planning and water harvesting.

CONCLUSION

There is a possibility that there will be no rainfall in June, July and October;

Forecasts for the months of May, June and July could be below 20 mm.

REFERENCES

LOPES, J. R. F.; SILVA, D. F. Application of the Mann-Kendall test to analyse rainfall trends in the state of Ceará. **Revista de Geografia,** v. 30, n. 3, 2013.

PATRIOTA, T.B. Archaeology in Cariri Paraibano. An Archaeological and Heritage Rescue in the Municipality of Camalaú. Revista tarairiú,v.l, n.7, p.80-92, 2014.

GALVÍNICIO, J.D.; SOUSA, F.A.S.; SHIRINIVASAM, V.S. Relief analysis of the Epitácio Pessoa Dam Hydrographic Basin. Geographical Magazine, p.54-69,2008

MENEZES, H. E. A.; BRITO, J. I. B.; SANTOS, C. A. C.; SILVA, L. L. The Relationship Between Tropical Ocean Surface Temperature and the Duration of Veranicos in the State of Paraíba. **Revista Brasileira de Meteorologia**, São José dos Campos, v.23, n.2, p.152-161, 2008.

MOURA, M. S. B.; GALVÍNCIO, J. D.; BRITO, L. T. L.; SOUZA, L. S. B.; SÁ, 1.1. S.; SILVA, T. G. F. **Climate and rainwater in the semi-arid region.** In: BRITO, L. T. L.; MOURA, M. S. B.; GAMA, G. F. B. (Org.). Potencialidades da água de chuva no Semi-Árido Brasileiro. 1 ed. Petrolina: Embrapa Semi-Arido, v. 1, p. 37-59, 2007.

OLIVEIRA, M. R. G.; CUNHA FILHO, M.; OLIVEIRA, E. P.; SERAFIM, M. V. A.; Statistical analysis of the precipitation time series in the municipality of São João do Cariri-PB. **Revista da Estatística,** v. 3, p.

144-148, 2014. Special edition: 59ª Annual Meeting of the Brazilian Branch of the International Biometric Society - RBRAS.

ROUCOU, P.; ARAGÃO, J. O. R.; HARZALLAH, A.; FONTAINE, B.; JANICOT, S. Vertical motion, changes to Northeast Brazil rainfall variability: A GCM simulation. **International Journal of Climatology**, West Sussex, v.16, n.l, p.879-891,1996.

SILVA, B. B.; LOPES, G. M.; AZEVEDO, P. V. Radiation balance in irrigated areas using landsat 5 - tm images. **Revista Brasileira de Meteorologia,** v.20, n.2, 243-252, 2005.

SILVA, I. S. M.; GUIMARAES, E. C.; TAVARES, M. Forecasting the average monthly temperature of Uberlândia, MG, with time series models. **Revista Brasileira de Engenharia Agrícola e Ambiental,** Campina Grande, vol.12, n.5, p. 1807-1929, 2008.

CHAPTER 4

RAINFALL FLUCTUATION IN THE SEMI-ARID REGION OF PARAIBÃO: WATER INSUFFICIENCY AND WATER SUPPLY

Dihego de Souza Pessoa, Julia Soares Pereira, Lilian Queiroz Firmino, Thalis Leandro Bezerra de Lima, Viviane Farias Silva and Vera Lucia Antunes de Lima

SUMMARY

Due to the irregularity of rainfall and low rainfall rates (below 800 mm per year), a large part of Brazil's semi-arid region faces an already chronic problem of water shortage, which is the reason for these obstacles to the development of agricultural and farming activities. The lack of efficient systems for storing water - a resource that is almost always concentrated in the hands of a few - further intensifies the social effects. Cycles of severe droughts, droughts and floods usually hit the region at intervals ranging from a few years to decades, as they help to disrupt the already fragile living conditions of the population living in the semi-arid region, particularly small producers and poor communities. The research was carried out in the municipality of Fagundes, in the state of Paraíba, which is part of the Borborema Plateau geoenvironmental unit, with altitudes ranging from 500 to 800 metres. The rainfall data was selected from the database of the Paraíba State Executive Water Management Agency (AESA), from 2000 to 2017, from the rainfall station installed in the town, and it was analysed that in 2014 the rainfall index was the highest.

(effective), with a downward trend in rainfall over the years. In relation to total climatology, the years 2000, 2004, 2009 and 2011 had high rainfall rates of over 1000 mm, and the year 2006 had the lowest total rainfall, with a 77.8% chance of a water deficit occurring in this region in the years studied. The function of rainfall *is* increasing, so as rainfall increases, the climatology tends to rise. The months from January to August are the most suitable for storing rainwater for various uses.

Keyword: drought; climatology; water storage

INTRODUCTION

Tropical regions have a rainfall regime known for its high rainfall intensity and prolonged duration, when natural phenomena such as drought do not occur and modify the predicted and expected behaviour. Semi-arid regions, on the other hand, are known for their low rainfall and desert-like climate. The variability of climate and rainfall in a region plays a major role in influencing how regional peoples and groups organise themselves in their social and economic work, including animal husbandry. As the climate is made up of a sum of integrated factors that determine the maintenance of life on the planet, it plays an important role in the behaviour of the climatic seasons, as it can either make it easier or more difficult for people to stay and establish themselves with their farming, industrial and any other activity that is present in any region of the globe (SLEIMAN, 2008).

According to Araújo et al. (2009), the irregular occurrence of rainfall makes it necessary to monitor and verify its behaviour by analysing the rainfall observed through strategically installed observation points, preferably those that are representative of a wider city or region. In this way, it is possible to keep track of its behaviour, which periods are considered to be rainy and which dry, the occurrence of abnormal phenomena, thus organising information by sectors and time periods, so that local climatology can be characterised with real values and the way the climate of a certain biome behaves in relation to neighbouring cities and even between different regions of the same state.

Monitoring a region's rainfall regime shows us that the scarcity of water, characteristic of the semi-arid region by nature, worsens local, social and economic problems every time it is verified that the water balance of what is precipitated and what is evapotranspirated and lost in any way is always negative, with total rainfall in general most of the time below the expected and observed average and never compensating above that average (MARENGO E SILVA DIAS, 2006). Analysing rainfall time series is an excellent planning and research mechanism for various areas, especially agriculture, as it always allows us to predict the benefits and losses that rainfall or the lack of it can bring (COSTA, BECKER and BRITO, 2013).

The state of Paraíba, with a large part of its territory in the semi-arid region, has a rainfall regime that is often compromised by long, hard periods of drought and lack of rainfall, which affects the way agricultural and animal husbandry activities are carried out, the latter being one of the most important in the industrial product processing scenario today to supply both the domestic and foreign markets with beef, pork, chicken, etc. Still on this path, when rainfall is considered intense and usually occurs without much meteorological forecasting, the drainage and run-off systems of cities and even degraded regions are jeopardised, while the lack of rain damages the level of reservoirs and affects the dynamics of production in the countryside (BECKER, MELO and COSTA, 2013).

With this in mind, the study was carried out with the aim of evaluating rainfall instability in the semi-arid region of Paraíba, highlighting the times of water scarcity and the times when water is likely to be used.

MATERIAL AND METHODS

The research was carried out in the municipality of Fagundes, in the state of Paraíba, which, according to CPRM (2005), is part of the Borborema Plateau geoenvironmental unit, with altitudes ranging from 500 to 800m, with an area of 162 km^2 , located at the following geographical coordinates: 7°20'45.56"S, 35°47'51.13"W (NOBRREGA et al., 2011). According to IBGE (2010), the municipality has approximately 11,830 inhabitants (FIGURE 1).

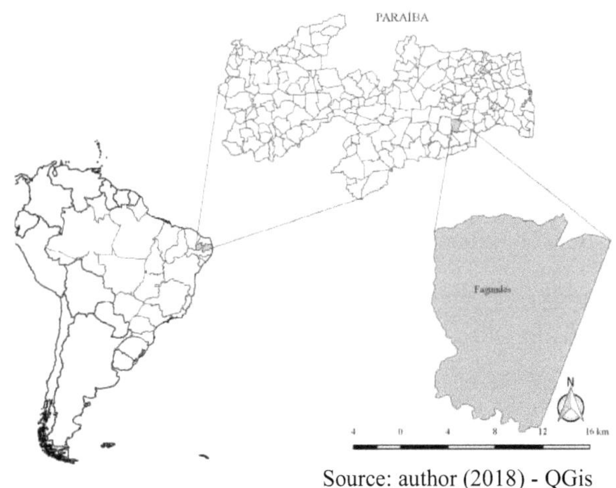

Source: author (2018) - QGis

Figura 1. Geographical location of the municipality of Fagundes-PB.

The rainfall data was selected from the database of the Paraíba State Executive Water Management Agency (AESA), from 2000 to 2017, from the rainfall station in Fagundes/PB. They were organised in an Excel spreadsheet and an analysis of monthly and annual rainfall was carried out, checking the maximum rainfall and

minimum, monthly, annual and the probability of precipitation above and below the 18-year average. The probability was calculated according to the methodology used by Bezerra et al. (2017). The programme Gretl (2018) was also used to estimate the probability using ordinary least squares (OLS) models.

RESULTS AND DISCUSSION

Figure 2 shows that only in January and June were the monthly averages higher than the climatological averages, with a maximum rainfall of 394 mm (January) and a minimum of 0 mm in several months. After June, there was a drop in rainfall, with the lowest averages in October and November. It can be seen that 2014 had the highest rainfall index (effective) but with a tendency for the rainfall rate to decrease over the course of the year.

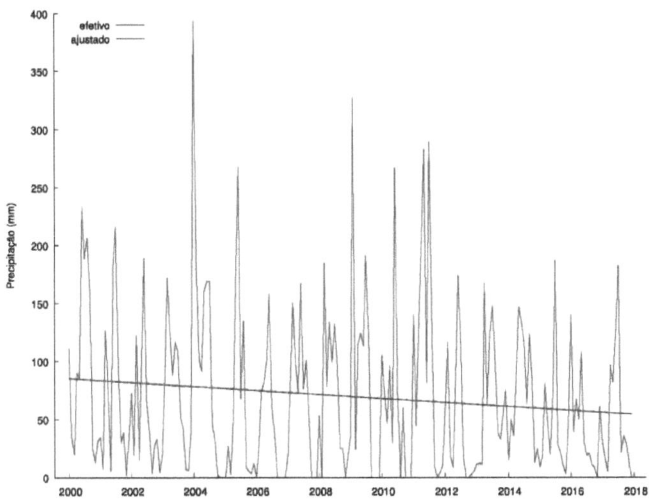

Figura 2. Rainfall time series from 2000 to 2017 in the municipality of Fagundes-PB.

Bezerra et al. (2017) analysed rainfall data in the municipality of Camalaú-PB and found that there were months without any rainfall records and that the highest record was in the month of April in 2014, with 223.3 mm. Alves et al. (2015) found that April and May had the highest volume of rainfall.

The total annual rainfall in relation to the total climatology can be seen in Figure 3, which shows that the years 2000, 2004, 2009 and 2011 had high rainfall indices above lOOmm, and the year 2006 had the lowest total rainfall, with a 77.8% chance of a water deficit occurring in this region in the years studied.

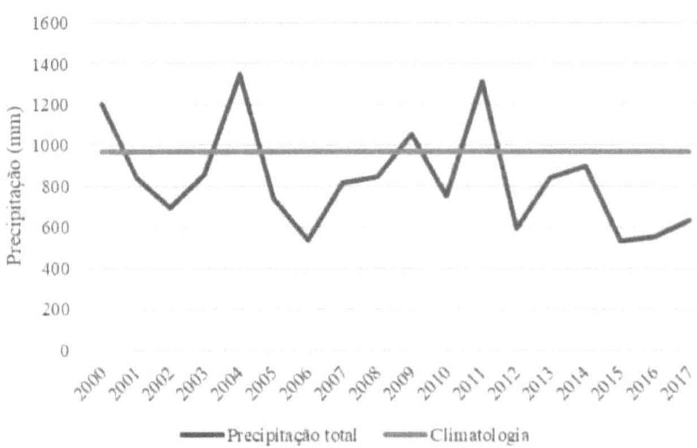

Figura 3. Total annual precipitation compared to annual climatology in the municipality of Fagundes-PB.

The probability of rainfall in January, June, July and August being below the climatological average mm is 55.5%, but there is a 44.4% chance of it being exceeded. In March, April and November, the chances of rainfall being lower than expected are around 89%. Above 60% of rainfall is reduced in the months of February, May, September, October and December. The months of January to August are therefore ideal for collecting and storing rainwater for consumption, irrigation and animals.

This shows that the climatology is superior to the rainfall data obtained, as can be seen in Figure 3, with the prospect of being exceeded by 22.22%.

Figure 4 shows that the climatology as a function of rainfall from 2000 to 2017 is increasing, so as rainfall increases, the tendency is for the climatology to increase. The climatology is based on the region's historical rainfall series, but since rainfall in the semi-arid region fluctuates and is due to periods of drought, there is irregularity, resulting in estimates that cannot be reached since the amount of local rainfall cannot be exact.

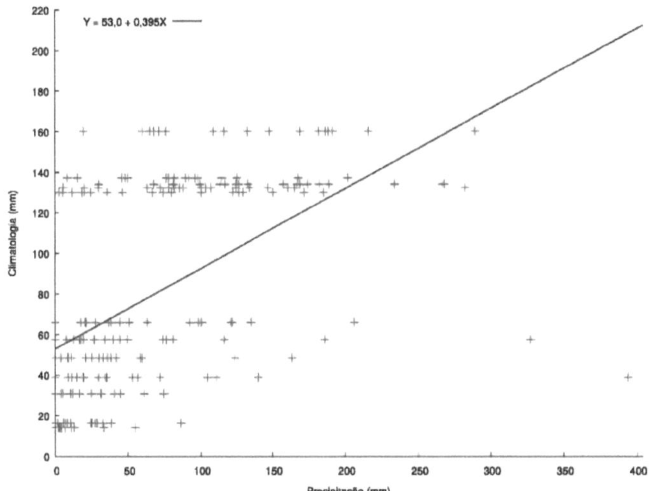

Figura 4. Time series of monthly climatology as a function of monthly precipitation in the municipality of Fagundes-PB.

Analysing the oscillation of rainfall in the municipality of Boqueirão, Silva et al. (2016) found that the probability of rainfall above 600 mm is 13.3%, while annual rainfall above 400 mm is 66.6% over a 15-year period. Bezerra et al. (2017), observing the temporal variation in the municipality of Camalaú-PB, realised that there is a 100% chance of rainfall being lower than expected.

CONCLUSION

Rainfall is below average, with a high probability of little precipitation;

The months of January to August are the most suitable for storing rainwater for various uses;

A 77.8 per cent chance of water deficit.

REFERENCES

ALVES, M.F.A.; ALVES, L.R.A.; SARMENTO, E.B.; LIMA, G.A.; CRISPIM, D.L. Analysis of rainfall in Pombal-PB related to active atmospheric systems. Revista verde de agroecologia e desenvolvimento sustentável,v. 10, n.2, p.169-175, 2015.

ARAÚJO, L. E. de. MORAES NETO, J. M. de. SOUSA, F. de A. S. de. Climate analysis of the Paraíba river basin - Rainfall Anomaly Index (IAC). **Environmental**

Engineering. Espirito Santo do Pinhal: v. 6, n. 3, p. 508-523. set/dez. 2009.

BECKER, C. T.; MELO, M. M. S.; COSTA, M. N. M. Temporal performance of rainfall series in the state of Paraíba: a comparative analysis. In: International Workshop on Water in the Brazilian Semi-Arid, 1, 2013, Campina Grande. **Proceedings...** 1st International Workshop on Water in the Brazilian Semi-Arid. Campina Grande: REALIZE, 2013. v. l,p.l-5.

BEZERRA, C.V.C.; SILVA, V.F.; BRITO, K.S.A.; PEREIRA, J.S.; LIMA, V.L.A. Semi-arid climatology of the municipality of Camalaú-PB. In: III Simposio Brasileiro de Recursos Naturais do Semiárido, Fortaleza- Ceará, 2017.

COSTA, M. N. M.; BECKER, C. T.; BRITO, J. I. B. Analysis of precipitation time series in the semi-arid region of Paraiba over a 100-year period - 1911 to 2010. **Revista Brasileira de Geografia Física,** v. 6, n. 4, p. 680-696, 2013.

CPRM - Geological Survey of Brazil. Registration of groundwater supply sources project. Diagnosis of the municipality of Fagundes, state of Paraíba / Organised [by] João de Castro Mascarenhas, Breno Augusto Beltrão, Luiz Carlos de Souza Júnior, Franklin de Morais, Vanildo Almeida Mendes, Jorge Luiz Fortunato de Miranda. Recife: CPRM/PRODEEM, 2005. 10 p.

IBGE (Brazilian Institute of Geography and Statistics) 2010. Available at http://www.ibge.gov .br/cidadesat/painel/painel.php?codmun=250610; accessed 7 Mar 2018.

MARENGO, J.; SILVA DIAS, P. Global climate change and its impacts on water resources. In: **Brazil's Fresh Waters: Ecological Capital, Use and Conservation.** São Paulo: Editora Escrituras, 2006. p. 63- 109. SILVA, V.F.; NASCIMENTO, E.C.; ANDRADE, L.O.; LIMA, V.L.A.; BARACUHY, J.G.V. Variabilidade pluviométrica no Municipio de Boqueirão-PB. In:II International Workshop on Water in the Brazilian Semi-Arid, 2016.

SLEIMAN, J, SILVA, M. E. S. **Precipitation Climatology and the Occurrence of Veranicos in the Northwestern Portion of the State of Rio Grande do Sul.** SIMPGEO/SP, Rio Claro, 2008.

CHAPTER 5

ANALYSIS OF RAINFALL OSCILLATIONS IN THE MUNICIPALITY OF BOQUEIRÃO IN THE SEMI-ARID REGION OF PARAIBANO

Julia Soares Pereira, Thalis Leandro Bezerra de Lima, Dihego Souza Pessoa, Lilian de Queiroz Firmino, Viviane Farias Silva and Vera Lucia Antunes de Lima

SUMMARY

The marked inter-annual variability of rainfall, associated with the low annual rainfall totals over the Northeast region of Brazil, *is* one of the main factors behind the occurrence of "droughts", characterised by a marked reduction in the seasonal rainfall total during the rainy season. The interannual variability of rainfall in this region is associated with variations in Sea Surface Temperature (SST) patterns over the tropical oceans, which affect the position and intensity of the Intertropical Convergence Zone (ITCZ) over the Atlantic Ocean. Under these characteristics, the research was carried out for the municipality of Boqueirão, which is located in the Boqueirão Microregion and the Borborema Mesoregion of the state of Paraíba, with the aim of analysing the probability of minimum, average and maximum rainfall in the municipality. The rainfall data was selected from the database of the Paraíba State Executive Water Management Agency (AESA), from 2000 to 2014 at the Boqueirão station/Boqueirão reservoir and was organised in an Excel spreadsheet and the probability of rainfall above 400 mm and rainfall below 400 mm was calculated. Irregularities were observed in the years studied, with the highest rate of rainfall in 2004 with 883.2 mm, while in 2012 there was the lowest occurrence of rain with 273.4 mm. The average for the period analysed was 483.54 mm, lower than the average suggested by the Department of Works Against Droughts (DNOCS), which is 661 mm. When analysing the probability of precipitation exceeding 400 mm, it showed 66.6%, while for values below 400 mm the possibility of occurrence is 33.3%, however the maximum and minimum precipitation can affect 6.6% in a time span of 14 years.

Keywords: Frequency of rainfall; Droughts; Probability

INTRODUCTION

The Northeast covers 18.27 % of Brazil's territory, with an area of 1,561,177.8 km^2 ; of these, 962,857.3 km^2 fall within the so-called Polígono das Secas (Drought Polygon), delimited in 1936 and revised in 1951, of which 841,260.9 km^2 cover the Northeastern semi-arid region. This shows that the territorial area of the semi-arid region, thus delimited, was larger than the sum of the territories of Germany, Italy, Cuba and Costa Rica (ARAÚJO, 2011). With the highest evaporation rates in Brazil, the semi-arid region of the Northeast, due to the high incidence of sunshine, with rates of approximately 2200 mm/year (ROCHA and KURTZ, 2001), has difficulties in water availability.

Marengo (2008) mentions the serious repercussions caused by the phenomenon of drought in semi-arid regions, noting that water is a critical factor for local populations. According to the CPRM (2005), the polygon of droughts has a rainfall regime characterised by extreme irregularity in time and space. In this scenario, water scarcity is a major obstacle to socio-economic development and even the population's livelihood. The cyclical occurrence of droughts and their catastrophic effects are well known and date back to the dawn of Brazil's history. It is recognised that the variations in meteorological elements over the years determine the climatic characteristics of a region, in such a way that its socio-economic structure and even its means of production are dependent on these characteristics (MARIN et al., 2000).

This variability can affect the economic and social life of the population in various ways, such as energy generation, agricultural activities, industry and the entire productive sector (BRITTO et al., 2008). The variability of rainfall in the north-east of Brazil, especially in the semi-arid region, has already been the focus of several studies, including Hastenrath and Heller (1977), Moura and Shukla (1981), Kousky and Cavalcanti (1984) and Harzallah et al. (1996). Among the various factors that condition the rainfall regime in the semi-arid northeast, we have the presence of meteorological systems on different scales of time and space, which in turn play a fundamental role in the quality of the rainy season.

In this context, the research was carried out with the aim of analysing the probability of minimum, average and maximum rainfall in the municipality of Boqueirão in the state of Paraíba.

MATERIAL AND METHODS

The research was carried out for the municipality of Boqueirão, which is located in the Boqueirão Microregion and Borborema Mesoregion of the state of Paraíba. Its area is 425 km^2, representing 0.7524% of the state, 0.0273% of the region and 0.005% of the whole of Brazil. The seat of the municipality has an approximate altitude of 355 metres and is 146.0099 km from the capital (CPRM, 2005).

The rainfall data was selected from the database of the Paraíba State Executive Water Management Agency (AESA) over a 15-year period, from 2000 to 2014 at the Boqueirão station/Boqueirão reservoir. They were organised in an Excel spreadsheet and the probability of rainfall above 400 mm and rainfall below 400 mm was calculated, as well as the possibility of maximum and minimum rainfall.

The probability was calculated as follows. Given a sample space S, with n (S) elements, and an event a of S, with n(S) elements, the probability of event A is P(A) such that: P(A) = n(A)/n(S), where n(S) is the 15 years studied and n(A) the occurrence of rainfall above or below 400mm.

RESULTS AND DISCUSSION

Looking at the annual rainfall data (Figure 1) shows the irregularity of rainfall over the 15-year period.

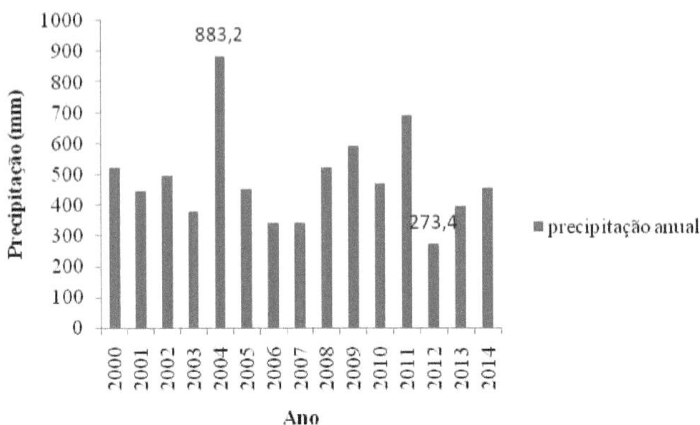

Figura 1. Annual rainfall variation in the municipality of Boqueirão.

As can be seen in Figure 1, the highest rainfall rate occurred in 2004 with 883.2 mm, while 2012 saw the lowest rainfall with 273.4 mm.

The average for the period analysed was 483.54 mm, which is below the average suggested by the Department of Works against Droughts (DNOCS) of 661 mm (DNOCS, 2015). The chance of the average of 661 mm or more occurring is 13.3 per cent over a 15-year period. When analysing the probability of precipitation above 400 mm, it is 66.6%, while for values below 400 mm the possibility of occurrence is 33.3%. However, the maximum and minimum rainfall that can occur is 6.6% over a 15-year period.

The frequency of positive and negative rainfall deviations in the periods of the historical series studied is shown in Figure 2. The annual rainfall deviations were quite variable and mostly negative, as a result of the irregular rainfall that occurs in this region. The years with the highest peaks are 2004, which was positive, and 2006, 2007, 2012 and 2013, which were negative.

Figura 2. Variation in the annual rainfall deviation in the municipality of Boqueirão.

Analysing the deviations, the years 2001, 2002, 2003, 2005, 2006, 2007, 2010, 2012, 2013 and 2014 can be considered dry years, while 2000, 2004, 2008, 2009 and 2011 generally had rainfall above 500 mm. Rainy years (2000, 2004, 2008, 2009 and 2011) can be considered those with normalised deviations above 12%.

Nascimento and Alves (2008), studying the ecoclimatology of the Cariri region of Paraíba, found that the rainfall regime combined with other physical factors, such as the low permeability of soils derived from crystalline rocks, has profound repercussions on the hydrological regime: the regime of the Cariris rivers is torrential and temporary, with violent floods, accelerated recession, zero discharge during most of the dry season, and even absent runoff during some years. Meanwhile, the dry season is characterised by a gradual reduction in river waters until they dry up, while floods occur abruptly. This scenario is witnessed throughout the semi-arid region, where droughts occur due to long dry spells.

CONCLUSION

The probability of an average annual rainfall of 661 mm *is* 13.3 per cent, while the chances of annual rainfall above 400 mm are 66.6 per cent and below 400 mm are 33.3 per cent, over a 15-year period. And the chance of maximum and minimum rainfall is 6.6%.

The municipality of Boqueirão is a region with irregular rainfall and the possibility of constant periods of drought.

REFERENCES

ARAÚJO, S.M.S. The semi-arid region of northeastern Brazil: Environmental issues and possibilities for sustainable use of resources. Rios Eletrónica- Revista Cientifica da FASETE. v.5, n.5, 2011.

BRITTO, P. F.; BARLETTA, R.; MENDONÇA, M. Spatial and temporal variability of rainfall in Rio Grande do Sul: influence of the EI Nino Southern Oscillation phenomenon. In: Brazilian Journal of Climatology. Year 4, August 2008. v. 3/4, p. 37-48.

CPRM. Geological Survey of Brazil. Project to register groundwater supply sources. Diagnosis of the municipality of Lagoa Seca, state of Paraíba. Recife: CPRM/PRODEM, 2005. Available at: <http://www.cprm.gov.br>. Accessed on: 25 July 2015.

DNOCS. National Department of Works Against Drought (DNOCS), 2015. Available at   :< http://www.dnocs.gov.br/~dnocs/php/comunicacao/monitoramento_de_res ervatorios.php>. Accessed on: 17 August 2015.

HARZALLAH A., ARAGÃO J. O. R., SADOURNY R.. Interannual Rainfall Variability in NorthEast Brazil: Observation and Model Simulation. International Journal of Climatology, 16, p. 816-176.1996.

HASTENRATH, S. , HELLER L.. Dynamics of Climatic Hazards in Northeast Brazil. Quarterly Journal Royal Meteorological Society, 103: (435). P. 77-92.1977

KOUSKY V. E., CAVALCANTI I. F. A.. Southern Oscillation - E1 Nino Event: Characteristics, Evolution and Precipitation Anomalies. Rev. Ciência e Cultura, 36: (11). São Paulo, p.18881899, 1984.

MARENGO, J. A. Water and climate change. Estud. av., São Paulo , v. 22, n. 63, p. 83-96, 2008 . Available at: <http://www.scielo.br/scielo.php?script=sci_arttext&pid=S0103401420080 00200006&lng=en&nrm=iso>. Accessed on: 17 August 2015. http://dx.doi.org/10.1590/S0103-401420080002000Q6.

MARIN, F. R.; SENTELHAS, P. C.; VILLA NOVA, N. A. Influência dos fenômenos EI Nino e La Nin ano clima de Piracicaba, SP. Revista Brasileira de Meteorologia, v.15, n.l, 123-129, 2000.

MOURA A. D., SHUKLA J.. On the Dynamics of Droughts in Northeast Brazil: Observations, Theory and Numerical Experiments with a General Circulation Model. Journal of the Atmospheric Sciences, 38: (12). p.2653-2675.1981.

NASCIMENTO, S.S.;ALVES,J.J.A. ECOCLIMATOLOGIA DO CARIRI PARAIBANO. Rev. Geogr. Acadêmica v.2 n.3, p. 28-41, 2008.

ROCHA, J. S. M; KURTZ, S. J. M. Integrated management of river basins. 4ª Edition. Santa Maria: UFSM, 2001. 302p.

I want morebooks!

Buy your books fast and straightforward online - at one of world's fastest growing online book stores! Environmentally sound due to Print-on-Demand technologies.

Buy your books online at
www.morebooks.shop

Kaufen Sie Ihre Bücher schnell und unkompliziert online – auf einer der am schnellsten wachsenden Buchhandelsplattformen weltweit! Dank Print-On-Demand umwelt- und ressourcenschonend produziert.

Bücher schneller online kaufen
www.morebooks.shop

Printed by Books on Demand GmbH, Norderstedt / Germany